Principles and Standards for School Mathematics

Navigating
through
Algebra
in
Grades 9–12

Maurice Burke
David Erickson
Johnny W. Lott
Mindy Obert

Johnny W. Lott
Grades 9–12 Editor

Peggy A. House
Navigations Series Editor

NCTM
NATIONAL COUNCIL OF
TEACHERS OF MATHEMATICS

Copyright © 2001 by
The National Council of Teachers of Mathematics, Inc.
1906 Association Drive, Reston, VA 20191-1502
(703) 620-9840; (800) 235-7566; www.nctm.org

All rights reserved

Second printing 2002

Library of Congress Cataloging-in-Publication Data:

Navigating through algebra in grades 9–12 / Maurice Burke ... [et al.].
    p. cm. — (Principles and standards for school mathematics navigations series)
  Includes bibliographical references.
  ISBN 0-87353-502-2
    1. Algebra—Study and teaching (Secondary) I. Burke, Maurice Joseph. II. Series.

QA159 .N387 2001
512'.007'2—dc21

2001030495

    Permission to photocopy limited material from *Navigating through Algebra in Grades 9–12* is granted for educational purposes. Permission must be obtained when content from this publication is used commercially, when the material is quoted in advertising, when portions are used in other publications, or when charges for copies are made. The use of material from *Navigating through Algebra in Grades 9–12*, other than in those cases described, should be brought to the attention of the National Council of Teachers of Mathematics.
    The publications of the National Council of Teachers of Mathematics present a variety of viewpoints. The views expressed or implied in this publication, unless otherwise noted, should not be interpreted as official positions of the Council.

Printed in the United States of America

Table of Contents

About This Book v

Introduction 1

Chapter 1
Algebra—the Language of Process 7
   What's the Area? 8

Chapter 2
Expanding the Notion of Variable 13
   When a Variable Is Not Single 16
   Tolerating a Bit 18
   Tolerance and Accuracy 19

Chapter 3
Expanding the Notion of Function Representation 23
   Would You Work for Me? 26
   The Devil and Daniel Webster 27
   Iterating to Find the Square Root of 2 28

Chapter 4
Expanding Students' Understanding of Algebraic Equivalence or Identity 31
   Representing the Solution Process by Graphing 34
   Families of Functions 38
   Operating on Functions 40

Chapter 5
Expanding Students' Understanding of Change 41

Chapter 6
Conclusion 47

Appendix
Blackline Masters and Solutions 49
   What's the Area? 50
   When a Variable Is Not Single 52
   Tolerating a Bit 56
   Tolerance and Accuracy 59
   Would You Work for Me? 63
   The Devil and Daniel Webster 66
   Iterating to Find the Square Root of 2 69
   Representing the Solution Process by Graphing 72
   Families of Functions 76
   Operating on Functions 80

References 85

CONTENTS OF CD-ROM

Applets
- General Grapher
- System of Equations
- Finite Differences

Blackline Masters

Readings from Publications of the National Council of Teachers of Mathematics

Design Your Own City: A Discrete Mathematics Project for High School Students
  Carol A. Bouma
  Discrete Mathematics across the Curriculum, K–12

Mapping Diagrams and the Graph of $y = \sin 1/x$
  Thomas J. Brieske
  Mathematics Teacher

Establishing Fundamental Concepts through Numerical Problem Solving
  Franklin Demana and Joan Leitzel
  The Ideas of Algebra, K–12

Mission Mathematics, Grades 9–12
  Peggy A. House

The Many Uses of Algebraic Variables
  Randolph A. Philipp
  Algebraic Thinking, Grades K–12: Readings from NCTM'S School-Based Journals and Other Publications

Some Misconceptions concerning the Concept of Variable
  Peter Rosnick
  Algebraic Thinking, Grades K–12: Readings from NCTM'S School-Based Journals and Other Publications

On the Meaning of Variable
  Alan H. Schoenfeld and Abraham Arcavi
  Algebraic Thinking, Grades K–12: Readings from NCTM'S School-Based Journals and Other Publications

Talking about Math Talk
  Miriam G. Sherin, Edith P. Mendez, and David A. Louis
  Learning Mathematics for a New Century

Hammurabi's Calculator
  Clifford Wagner
  The Teaching and Learning of Algorithms in School Mathematics

What Are These Things Called Variables?
  Sigrid Wagner
  Algebraic Thinking, Grades K–12: Readings from NCTM'S School-Based Journals and Other Publications

Solving Equations in a Technological Environment
  Michal Yerushalmy and Shoshana Gilead
  Algebraic Thinking, Grades K–12: Readings from NCTM'S School-Based Journals and Other Publications

About This Book

Principles and Standards for School Mathematics (National Council of Teachers of Mathematics [NCTM] 2000) outlines much of what has traditionally been algebra 1 in secondary schools as expected content for the middle grades. It is imperative then that a broadening and deepening of mathematics content take place in high school. New topics—not frequently before found in algebra—such as recursion, classes of functions, and using technology on symbolic expressions, are emerging in the high school curriculum.

As the mathematics broadens, deepens, and emerges, we must not only tie the new learning to what has been learned before but also do so in a way that helps students generalize what they have previously learned in a more specific way. Only in this manner can we approach algebra as a language of process for all students that adds a structural framework to the mathematics begun as arithmetic in the lower grades.

Algebra is an extremely effective tool for investigating all areas of school mathematics. As students deepen their understanding of number, operation, measurement, geometry, statistics, and probability, algebraic ideas constantly emerge. These emerging ideas give teachers many opportunities for focusing, reinforcing, and developing algebraic processes. The basic elements of algebraic processes are variables, operations, and relations, including functions. This book addresses algebra as a language of process in chapter 1 and expands the notion of variable in chapter 2. Chapter 3 develops the notion of the representation of functions, and chapter 4 extends students' understandings of algebraic equivalence. Chapter 5 helps expand students' understanding of change.

Each chapter is built around a set of activities that can be used either with students or for the professional development of teachers. The activities include lists of prerequisites and the materials to be used, including the blackline masters, which are signaled by an icon and can be found at the back of this book and on the accompanying CD-ROM. In addition, the CD, also signaled by an icon, contains readings for professional development, along with several tools, listed as applets on the contents page. The authors originally intended to include assessments for each activity in the book but decided against it. Instead of developing an assessment rubric unique to each student activity, we believe that teachers need to develop simple, more general rubrics that they can use on a regular basis, rubrics that translate into habits of mind for their students. For example, in promoting constructive discourse during an algebra activity, teachers can assess the quality of the dialogue by using the rubric "Explain, build, or go beyond," described in "Talking about Math Talk" (Sherin, Mendez, and Louis 2000, pp. 188–96). With this simple rubric, teachers can encourage students to justify or explain their procedures and results. Students are asked to build on one another's ideas and thereby, for example, contribute alternative procedures and ways of validating results. Finally, the rubric encourages students to engage routinely in generalizing beyond the specifics of the problem or process under study.

In activities in which building an understanding of an algebraic

Key to Icons

Principles and Standards

CD-ROM

Blackline Master

Three different icons appear in the book, as shown in the key. One alerts readers to material quoted from *Principles and Standards for School Mathematics,* another points them to supplementary materials on the CD-ROM that accompanies the book, and a third signals the blackline masters and indicates their locations in the appendix.

process is the goal, another simple rubric such as "Estimate, carry out, check, think back" might be used. Using this rubric, teachers routinely ask such questions as What would be a good estimate? Why do you think your method works? What are some ways you could check your answer? How could you have done that another way? What other problems would your method help in solving? It is important that students occasionally write down their responses to these questions so they can be used in assessing their understanding. These and other such questions are important in environments where students are making regular use of technology to carry out algebraic procedures. However, they are also helpful in environments where a computer algebra system (CAS)—which enables the user to work with variables on a calculator—is not available.

In *Navigating through Algebra in Grades 9–12*, the authors have tried to produce a book that will cause mathematics teachers to think. Collectively, we believe that mathematics education in the United States will reflect the notions in *Principles and Standards for School Mathematics* (NCTM 2000) only when teachers' beliefs are both validated and challenged. In the past, *algebra* evoked in almost everyone a similar image—that of a traditional ninth-grade course from the 1960s, 1970s, or 1980s. The educational mindset reflected in that image began to change with the publication of *Curriculum and Evaluation Standards for School Mathematics* (NCTM 1989). *Principles and Standards* is again challenging that mindset with recommendations for algebraic thinking reaching from prekindergarten to grade 12. With a challenged mindset come new topics and old topics thought of in different ways. With the emphasis on algebraic thinking from the prekindergarten years on, concepts and expectations in algebra have changed and will continue to change. The use of technology is now expected, not to the exclusion of algebraic computation by hand, but in ways that enhance how students learn and understand computation. As you read and work through this book, consider how your classroom must change to implement the Algebra Standard.

NAVIGATIONS SERIES

GRADES 9–12

NAVIGATING through ALGEBRA

Throughout history, algebra has been a cornerstone of mathematics, and we can trace the roots of algebraic thinking deep into the bedrock of mathematics. Thus it is not surprising that algebra has emerged as one of the central themes of *Principles and Standards for School Mathematics* (National Council of Teachers of Mathematics [NCTM] 2000), for algebra continues to be an essential component of contemporary mathematics and its applications in many fields. Yet in the school curriculum, algebra has been too often misunderstood and misrepresented as an abstract and difficult subject to be taught only to a subset of secondary school students who aspire to study advanced mathematics; in truth, algebra and algebraic thinking are fundamental to the basic education of all students, beginning in the earliest years.

Algebra is frequently described as "generalized arithmetic," and indeed, algebraic thinking is a natural extension of arithmetical thinking. Both arithmetic and algebra are useful for describing important relationships in the world. But although arithmetic is effective in describing static pictures of the world, algebra is dynamic and a necessary vehicle for describing a changing world. Even young children can appreciate the significance of change and the need to describe and predict variation. Algebraic thinking begins with the very young, expands and deepens and matures through the years, and continues to serve adults long after the end of formal schooling. To achieve that outcome requires an algebra curriculum that is coherent and developmental, that is anchored by important mathematical concepts, and that is well articulated and coordinated across the grades.

The Navigations series seeks to guide readers through the five strands of the *Principles and Standards for School Mathematics* in order to help them translate the Standards and Principles into action and to illustrate the growth and connectedness of content ideas from prekindergarten through grade 12. The Navigations through the algebra curriculum reflect NCTM's vision of how algebraic concepts should be introduced, how they grow, what to expect of students during and at the end of each grade band, how to assess what students know, and how selected instructional activities can contribute to learning.

Fundamental Components of Algebraic Thinking

The Algebra Standard emphasizes relationships among quantities and the ways in which quantities change relative to one another. To think algebraically, one must be able to understand patterns, relations, and functions; represent and analyze mathematical situations and structures using algebraic symbols; use mathematical models to represent and understand quantitative relationships; and analyze change in various contexts. Each of these basic components evolves as students grow and mature.

Understanding patterns, relations, and functions

Young children begin to explore patterns in the world around them through experiences with such things as color, size, shape, design, words, syllables, musical tones, rhythms, movements, and physical objects. They observe, describe, repeat, extend, compare, and create patterns; they sort, classify, and order objects according to various characteristics; they predict what comes next and identify missing elements in patterns; they learn to distinguish types of patterns, such as growing or repeating patterns.

In the higher elementary grades, children learn to represent patterns numerically, graphically, and symbolically, as well as verbally. They begin to look for relationships in numerical and geometric patterns and analyze how patterns grow or change. By using tables, charts, physical objects, and symbols, students make and explain generalizations about patterns and use relationships in patterns to make predictions.

Students in the middle grades explore patterns expressed in tables, graphs, words, or symbols, with an emphasis on patterns that exhibit linear relationships (constant rate of change). They learn to relate symbolic and graphical representations and develop an understanding of the significance of slope and y-intercept. They also explore "What if?" questions to investigate how patterns change, and they distinguish linear from nonlinear patterns.

In high school, students create and use tables, symbols, graphs, and verbal representations to generalize and analyze patterns, relations, and functions with increasing sophistication, and they convert flexibly among various representations. They compare and contrast situations modeled by different types of functions, and they develop an understanding of classes of functions, both linear and nonlinear, and their properties. Their understanding expands to include functions of more than one variable, and they learn to perform transformations such as composing and inverting commonly used functions.

Representing and analyzing mathematical situations and structures using algebraic symbols

Young children can illustrate mathematical properties (e.g., the commutativity of addition) with objects or specific numbers. They use objects, pictures, words, or symbols to represent mathematical ideas and relationships, including the relationship of equality, and to solve problems. When children are encouraged to describe and represent quantities in different ways, they learn to recognize equivalent representations and expand their ability to use symbols to communicate their ideas.

Later in the elementary grades, children investigate, represent, describe, and explain mathematical properties, and they begin to generalize relationships and to use them in computing with whole numbers. They develop notions of the idea and usefulness of variables, which they may express with a box, letter, or other symbol to signify the idea of a variable as a placeholder. They also learn to use variables to describe a rule that relates two quantities or to express relationships using equations.

During the middle grades, students encounter additional uses of variables as changing quantities in generalized patterns, formulas, identities, expressions of mathematical properties, equations, and inequalities. They explore notions of dependence and independence as variables change in relation to one another, and they develop facility in recognizing the equivalence of mathematical representations, which they can use to transform expressions; to solve problems; and to relate graphical, tabular, and symbolic representations. They also acquire greater facility with linear equations and demonstrate how the values of slope and y-intercept affect the line.

High school students continue to develop fluency with mathematical symbols and become proficient in operating on algebraic expressions in solving problems. Their facility with representation expands to include equations, inequalities, systems of equations, graphs, matrices, and functions, and they recognize and describe the advantages and disadvantages of various representations for a particular situation. Such facility with symbols and alternative representations enables them to analyze a mathematical situation, choose an appropriate model, select an appropriate solution method, and evaluate the plausibility of their solution.

Using mathematical models to represent and understand quantitative relationships

Very young children learn to use objects or pictures, and, eventually, symbols to enact stories or model situations that involve the addition or subtraction of whole numbers. As they progress into the upper elementary grades, children begin to realize that mathematics can be used to model numerical and geometric patterns, scientific experiments, and other physical situations, and they discover that mathematical models have the power to predict as well as to describe. As they employ graphs, tables, and equations to represent relationships and use their models to draw conclusions, students compare various models and investigate whether different models of a particular situation yield the same results.

Contextualized problems that can be modeled and solved using various representations, such as graphs, tables, and equations, engage middle-grades students. With the aid of technology, they learn to use functions to model patterns of change, including situations in which they generate and represent real data. Although the emphasis is on contexts that are modeled by linear relationships, students also explore examples of nonlinear relationships, and they use their models to develop and test conjectures.

High school students develop skill in identifying essential quantitative relationships in a situation and in determining the type of function with which to model the relationship. They use symbolic expressions to represent relationships arising from various contexts, including situations in which they generate and use data. Using their models, students conjecture about relationships, fomulate and test hypotheses, and draw conclusions about the situations being modeled.

Analyzing change in various contexts

From a very early age, children recognize examples of change in their environment and describe change in qualitative terms, such as getting taller, colder, darker, or heavier. By measuring and comparing quantities, children also learn to describe change quantitatively, such as in keeping track of variations in temperature or the growth of a classroom plant or pet. They learn that some changes are predictable but others are not and that often change can be described mathematically. Later in the elementary grades, children represent change in numerical, tabular, or graphical form, and they observe that patterns of change often involve more than one quantity, such as that the length of a spring increases as additional weights are hung from it. Students in the upper grades also begin to study differences in patterns of change and to compare changes that occur at a constant rate, such as the cost of buying various numbers of pencils at $0.20 each, with changes whose rates increase or decrease, such as the growth of a seedling.

Middle-grades students explore many examples of quantities that change and the graphs that represent those changes; they answer questions about the relationships represented in the graphs and learn to distinguish rate of change from accumulation (total amount of change). By varying parameters such as the rate of change, students can observe the corresponding changes in the graphs, equations, or tables of values of the relationships. High school students deepen this understanding of how mathematical quantities change and, in particular, of the concept of rate of change. They investigate numerous mathematical situations and real-world phenomena to analyze and make sense of changing relationships; interpret change and rates of change from graphical and numerical data; and use algebraic techniques and appropriate technology to develop and evaluate models of dynamic situations.

Developing an Algebra Curriculum

Clearly, an algebra curriculum that fosters the development of algebraic thinking described here and in *Principles and Standards* (NCTM 2000) must be coherent, focused, and well articulated. It cannot be merely a collection of lessons or activities but must instead be developmental and connected. Mathematical ideas introduced in the early years must deepen and expand, and subsequent instruction should link to, and build on, that foundation. As they progress through the curriculum, students must be continually challenged to learn and apply increasingly more-sophisticated algebraic thinking and to solve problems in a variety of school, home, and life settings.

The Navigations cannot detail a complete algebra curriculum, nor do they attempt to do so. Rather, the four Navigations through the Algebra Standard illustrate how a few selected "big ideas" of algebra should develop across the prekindergarten–grade 12 curriculum. Further, the topical strands of the mathematics curriculum, such as algebra, geometry, and data analysis, are highly interconnected, and many of the concepts presented under one strand will further develop and deepen when encountered again in another context. Thus, future Navigations through other parts of *Principles and Standards* (NCTM 2000) will reinforce the algebra objectives, and vice versa.

The methods and ideas of algebra are an indispensable component of mathematical literacy in contemporary life, and the algebra strand of the curriculum is central to the vision of mathematics education set forth in *Principles and Standards for School Mathematics*. These Navigations are offered as a guide to help educators set a course for successful implementation of the important Algebra Standard.

NAVIGATIONS SERIES

GRADES 9–12

NAVIGATING through ALGEBRA

Chapter 1
Algebra–the Language of Process

The General Grapher applet on the CD-ROM allows users to explore a variety of situations described in this book.

Algebra is used to describe mathematical and real-world processes and to study their properties. In this endeavor, algebra makes extensive use of the concepts of variable, operation, function, and relation. It exploits graphical features of processes as well as the "change" attributes of processes. In this chapter, we consider an example that manifests many of the aspects of algebra that *Principles and Standards for School Mathematics* (National Council of Teachers of Mathematics [NCTM] 2000) recommends be included in the curriculum. In the following example, algebra is used to model the process of computing area. That is, algebra is used to generate a function representing the generalization of the arithmetic process that students use. It is inherent in their solutions that the students have previously learned that the area formula for a rectangle yields positive values, is additive, and is computed by multiplying length by width. The systematic guessing strategy depends on first choosing what to vary (the key variable) in modeling the process. Depending on how systematic the guessing phase of data collection is (for example, guessing values for the width at equal intervals from 1 m to 50 m), the modeling of the process can also be approached by looking at the change features and by looking at the *recursive* features of the change process. Consider What's the Area?, an activity based on a classic problem.

What's the Area?

Goals

- Examine an algebra problem from different perspectives
- Find equivalent forms of equations
- Use multiple representations to examine a problem

Materials and Equipment

- One copy of the activity page for each student
- Spreadsheet software
- Computer algebra systems, if available

Activity

A subdivision is being placed on a piece of land 1000 m by 1500 m. A boulevard of trees and an access road of uniform width form the border of the subdivision, as shown in figure 1.1. The area of the inner rectangle of houses and parks is to be at least 1.35 million m² to accommodate the planned homes and parks. What is the largest width that can be set aside inside the perimeter for the border composed of the boulevard and road?

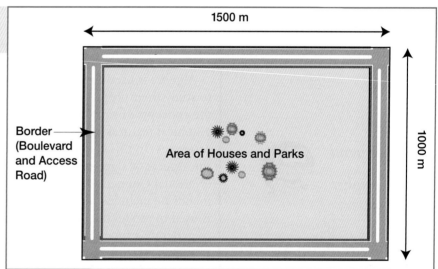

Fig. **1.1.**
A design for a subdivision

Discussion

Teachers can ask their students, or cooperative groups of students, to make systematic guesses for the width, starting with a discussion of a feasible domain for the guesses. (For example, it would make no sense to make any guesses larger than 500 m.) This method of using systematic guesses for a problem is outlined in Demana and Leitzel (1988) and offers teachers an investigative approach that can be applied to many problems. The students can contribute their results to a class data table (see fig. 1.2) and create a graph, as shown in figure 1.3.

Such estimate-check-reestimate approaches give everyone a role in the solution process while providing the teacher important assessment information about students' understanding. For example, students who make arithmetic errors, who leave something out in their analysis of the task, or who do not know how to compute areas will see that their results do not fit the broader pattern that emerges. Teachers who use cooperative learning groups can have students check one another's results before submitting them.

Width (Meters)	Area of Houses and Parks (Square Meters)
25	1,377,500
50	1,260,000
75	1,147,500
100	1,040,000
125	937,500
150	840,000
175	747,500
200	660,000
225	577,500
250	500,000
275	427,500
300	360,000
325	297,500
350	240,000
375	187,500
400	140,000
425	97,500
450	60,000
475	27,500
500	0

Fig. **1.2.**
Students' guesses and results

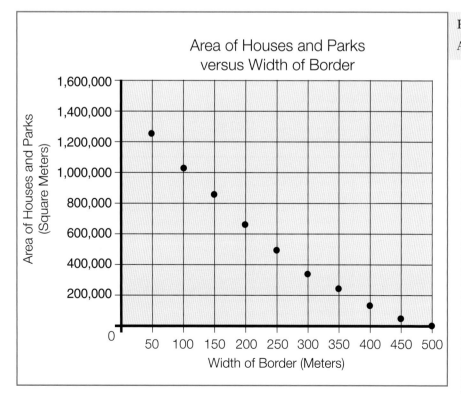

Fig. **1.3.**
A graph of the subdivision data

Chapter 1: Algebra—the Language of Process

At this stage, the students can fit the data with a function by discussing the process they used to check their initial guesses. See, for example, Miguel's (fig. 1.4) and Arrin's (fig.1.5) explanations of the processes they used to check their initial guesses. Note that both students' guesses leave too little space for the area for the houses and parks; they must reduce their guesses for the width of the border.

Fig. **1.4.**
Miguel's solution

I guessed 100 meters. That means the area of the road and boulevard is $100 \cdot 100 + 800 \cdot 100 + 100 \cdot 100 + 1300 \cdot 100 + 100 \cdot 100 + 800 \cdot 100 + 100 \cdot 100 + 1300 \cdot 100$.
I subtracted this from $1000 \cdot 1500$ to get the area left for the houses. My housing project is below.

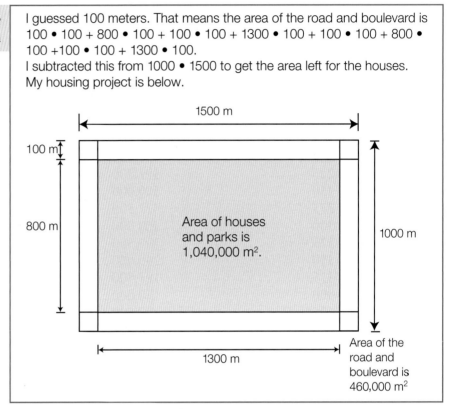

Fig. **1.5.**
Arrin's solution

I guessed 75 meters. I then computed the length and width of the inner rectangle to be length = $1000 - 2 \cdot 75$ and width = $1500 - 2 \cdot 75$. I multiplied them to get the area left for the houses.

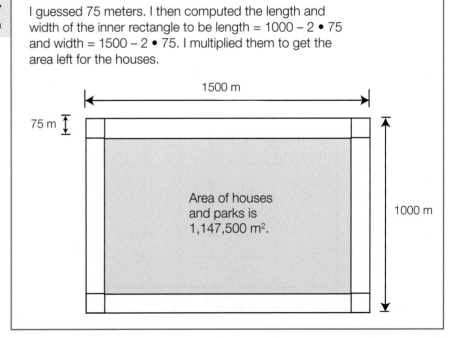

Multiple processes often emerge. These processes can be generalized by asking students to use their same approaches on an estimate the teacher made without telling the students what that estimate was. They can call the teacher's estimate g. In this fashion, functions representing the processes used by the students are generated. Miguel's process could be summarized as

$$Area\,(g) = 1500 \cdot 1000 - [g^2 + g(1000 - 2g) + g^2 + g(1500 - 2g)$$
$$+ g^2 + g(1000 - 2g) + g^2 + g(1500 - 2g)].$$

Arrin's process can be summarized by the function

$$Area\,(g) = (1500 - 2g)(1000 - 2g).$$

Although they may appear different, both Miguel's and Arrin's expressions are the same function. Identifying equivalencies is an important algebraic process. The identification process requires us to consider the implications of the two different processes—Miguel's, $m(x)$, and Arrin's, $a(x)$—in showing that they are algebraically equivalent. The challenge of showing the equivalence boils down to showing that the functions $m(x)$ and $a(x)$ form the identity $m(x) = a(x)$, possibly by using graphs, tables, and the basic algebraic properties of the operations used in the functions.

To deepen their understanding, the students might use a computer algebra system to change Miguel's and Arrin's functions to the same form, thereby showing that they are the same. The students might also discuss what it means for algebraic expressions to be equivalent. As students learn that algebra is a language for representing processes, they also realize that it provides tools for generating equivalent but alternative equations to solve a problem. Discussing why equations are equivalent helps a teacher focus attention on the algebraic process of multiplication and combining like terms. Discussions of how to check answers after symbolic manipulations have occurred lead to discussions of factoring, the distributive property, identities, and algebraic equivalence.

Solving equations can often be thought of as an algebraic specification process. In this example, this process involves imposing the condition that the area of the inner rectangle must be at least 1.35 million square meters, which leads to constraints on the x-values, the width of the border. This imposition leads students to either solving a quadratic equation symbolically or using graphical-approximation methods or table methods to solve the equation. It involves solving inequalities, which also can be done visually through graphs and numerically through tables.

At this point in the example, the students might generate an answer to the original question using the table to estimate an answer between 25 and 50 for the width and then using a finer increment to be more exact. Or they could use a graph and find the intersection of the line with equation $y = 1,350,000$ and the equation of the quadratic they developed. Finally, they could use the "solve" feature of the computer algebra system to find when the value of the area function equals 1.35 million square meters—for example, to solve $1,350,000 = (1500 - 2g)(1000 - 2g)$ for g in order to get $625 - 25\sqrt{565}$, or about 30.7 meters for the width of the border.

The "efficiency" of the various functions generated in the example

Chapter 1: Algebra–the Language of Process

Efficiency analyses can be made when students encounter the associative property of multiplication of matrices. For example, students might first discover that the product of an $N \times M$ matrix and an $M \times R$ matrix requires $N \times M \times R$ multiplications. Then they might compare $(A \times B) \times (C \times D)$ with $[A \times (B \times C)] \times D$, where the dimensions of A, B, C, and D are, respectively, 15×5, 5×100, 100×3, and 3×40. The first case requires 79,500 multiplications, whereas the second case requires only 3,525 multiplications. Both cases give the same matrix product because of the associative property of matrix multiplication.

above is a topic that has rarely been mentioned in algebra classes, but it provides an opportunity for teachers to open a class discussion. Such a discussion leads to a different view of the operations involved in the algebraic process represented by the functions. The expression being subtracted in Miguel's function is computationally of the form

$$a^2 + a(b-2a) + a^2 + a(c-2a) + a^2 + a(b-2a) + a^2 + a(c-2a),$$

where a is the width of the road and b and c are the dimensions of the rectangular subdivision, whereas Arrin's function is of the form

$$(c-2a)(b-2a).$$

In this case it is clear which method is computationally more efficient. But how much more efficient is Arrin's method than Miguel's? Once entered into a calculator, both functions appear to return values instantaneously for the area. But technology can be used to expand the exploration of the efficiency. If one enters large numbers for a, b, and c in the calculator in the computational forms above and performs the operation ten times in a row, Arrin's function requires about half the computing time that Miguel's requires.

A teacher can repeat this experiment to show that identities such as $a(b + c) = ab + ac$ or $(a + b)(c + d) = ac + ad + bc + bd$ are not equally time efficient on a calculator. Such excursions give a new sense to what it means to "simplify the expression."

Not included here, but a possible extension of the activity, is to think of algebra as generalizing the process represented by the functions $m(x)$ and $a(x)$. Students can be asked to examine the process they used in even more general terms where the area of the inner rectangle is to be some fixed but unspecified value, v. Or a different generalization might be to specify the dimensions of the subdivision only as l and w. By looking at their processes as being part of a broader family of processes, students are led to the examination of the family properties of functions.

Navigations Series

Grades 9–12

Navigating through Algebra

Chapter 2
Expanding the Notion of Variable

This list of the uses of variables is not unlike the list given by Usiskin (1988) in The Ideas of Algebra, K–12.

The example in chapter 1 illustrates how students can use the concept of variable in many different mathematical settings:

- In formulas such as that for the area of a rectangle
- In generalizing arithmetic processes
- In representing functions
- In statements of identities
- As unknowns in equations and inequalities to be solved
- In specifying parameters in descriptions of families of functions

In each of the uses, the variable stands for an arbitrary member of a set of numbers. *Principles and Standards for School Mathematics* (National Council of Teachers of Mathematics [NCTM] 2000) suggests that teachers extend the notion of variable so that students can see that it is a useful concept for studying processes whose objects are not represented by single numbers but perhaps by lists of numbers, intervals of numbers, or sets of points in a plane.

As an introduction to an expanded notion of variable for secondary schools, we must think about what new knowledge we expect students to acquire. In their middle-grades experiences and before, they may have considered variables only to represent single numbers. They should have worked with such patterns as that in the following problem:

Find a possible next term: 3, 6, 9, 12, 15, …

In the lower grades, students learned that a pattern for the sequence above was an "Add 3" pattern; they may or may not have been aware that the number of the term could be considered a variable in producing the terms of the sequence. In the middle grades, the "Add 3" pattern may have been generalized to the form $3n$, where n is a natural number (or the domain of n is the set of natural numbers). The students may have graphed the function $f(n) = 3n$ to discover that the sequence is linear, though discrete, as seen in figure 2.1.

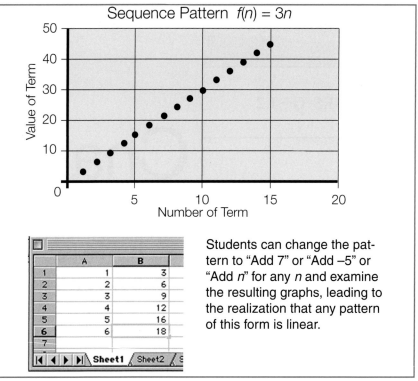

Fig. 2.1.

A sequence pattern $f(n) = 3n$, where n is a natural number. The graph of the "Add 3" pattern was constructed from the spreadsheet.

Students can change the pattern to "Add 7" or "Add –5" or "Add n" for any n and examine the resulting graphs, leading to the realization that any pattern of this form is linear.

The problem may be approached differently using column A as is; then to find the value in cell B1, use the explicit formula =A1*3. To complete column B, the fill-down command may be used by highlighting cell B1 and filling down. This operation basically finds a value for $3n$, where n is a natural number. Note that $f(n) = 3n$, where n is a natural number, is an explicit function that represents the recursive function NEXT = NOW + 3.

A spreadsheet was used to produce the values graphed in figure 2.1. If the number of the term of a sequence is in column A of a spreadsheet, then one way to find the terms of the sequence in column B is to use the first term as 3 with a recursive formula such as = B1 + 3 as the value in cell B2 and fill down. This type of recursive formula is the "natural formula" used by young children when they talk about an "Add 3" pattern. With the use of this recursive formula and the fill-down command, students could find as many terms of the sequence as desired. (The use of recursion is examined in more detail in chapter 3.)

Extensions

For grades 9–12, the "Add 3" pattern, or the recursive pattern of the spreadsheet, could be used as a springboard to expand the concept of variable and to set up the study of functions and variables in other settings and contexts. For example, *Principles and Standards* (NCTM 2000) lists as expectations for students in grades 9–12 the ability to understand and perform transformations (p. 296) and to use transformations to analyze mathematical situations (p. 308). In geometry, to consider a reflection in a line with equation $y = x$, we should be concerned about what happens to the entire plane, not just a single point. In other words, we are concerned with what happens to the set of all points with coordinates of the form (x, y), where x and y are real numbers (the domain of the function symbolized by $\Re \times \Re$). To examine such a

Following Principles and Standards (NCTM 2000, p. 304), the recursive relation might be seen as

$$NEXT = NOW + 3$$

or as

$$\begin{cases} y_1 = 3 \\ y_{n+1} = y_n + 3 \end{cases}.$$

14   Navigating through Algebra in Grades 9–12

reflection, we frequently focus on what happens to a set of three non-collinear points, *A*, *B*, and *C*, that are the vertices of triangle *ABC* with respective coordinates (0, 0), (1, 0), and (0, 1), as shown in figure 2.2.

What happens to all points of the plane can be determined by the images of three noncollinear points of the plane. Thus, we look at the reflection image ($r_{y=x}(\triangle ABC)$) of triangle *ABC* in the line. Here we consider the reflection, or function $r_{y=x}$, acting not on the entire plane (its domain) but on a subset, triangle *ABC*. If students have had no previous experience thinking about a function acting on more than a single input at a time, it can be difficult for them to adjust to thinking about a variable representing a plane or a subset of a plane. Typically, students think they are seeing a brand new concept never encountered before when they study transformations in geometry.

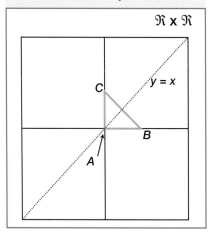

Fig. **2.2.**

A reflection over line $y = x$

Extensions that build on middle-grades algebra experiences should be provided in high school. For example, the function $f(n) = 3n$, where *n* is a natural number, may be reexamined using extensions in several areas. Visually we can think about the function with a mapping diagram, as in figure 2.3. The domain is pictured on the lower number line; the set of images is contained in the upper number line. The arrows indicate the pairing of *n* to 3*n*. The set of images of 1, 2, 3, 4, …, 50 contains the numbers 3, 6, 9, …, 150; symbolically, the situation can be expressed

$$f(\{1, 2, 3, …, 50\}) = \{3, 6, 9, …, 150\}.$$

Furthermore, the problem could be thought of algebraically as follows:

$$1 \leq n \leq 50,$$

where *n* is a natural number, and

$$3 \cdot 1 \leq 3 \cdot n \leq 3 \cdot 50$$
$$3 \leq 3n \leq 150.$$

In words, the inequalities indicate that for any natural number greater than or equal to 1 and less than or equal to 50, the images of these numbers are multiples of 3 from 3 to 150.

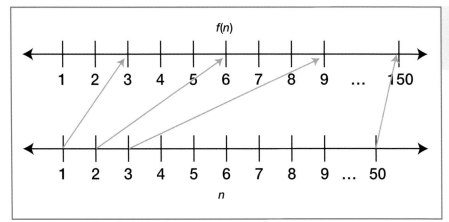

Fig. **2.3.**

A mapping diagram for $f(n) = 3n$, where *n* is a natural number

The following activity, When a Variable Is Not Single, is designed to allow students to work through the exercise using both mapping diagrams and a spreadsheet.

When a Variable Is Not Single

Goals

- Examine variables as a set of objects
- Find the image of a set of objects

Materials and Equipment

- A copy of the activity pages for each student
- Spreadsheet software

p. 52

Discussion of the Activity

In this activity, no longer does the spreadsheet user have to consider what happens when a single number is substituted for *n* in the function to gain an output; instead, the user may look at a series of values of *n* as an input for the fill-down command. The set of corresponding values of $f(n) = 3n$ are the images of these values. For example, one can think about all the images of all the natural numbers less than or equal to 50 under this function, as shown in the mapping diagram in figure 2.3.

An extension of the problem in the activity is to suppose that the user wants the multiples of 3 (in column B of the spreadsheet) greater than 100 but less than 1000. This subset is the set of images of some set of preimages from the domain. To determine this set, the user can think about what values in column A produce the desired values in column B. This determination can also be made by solving inequalities for *n*, as seen in most first-year algebra books. For example:

$$100 \leq f(n) \leq 1000$$
$$100 \leq 3n \leq 1000$$
$$\frac{100}{3} \leq \frac{3n}{3} \leq \frac{1000}{3}$$
$$33\frac{1}{3} \leq n \leq 333\frac{1}{3}$$

Since *n* is a natural number, we know that the bounds for *n* are 34 and 333.

On a spreadsheet, we might just look for the values near 100 and 1000 in column B and find the corresponding values in column A. Consider that the spreadsheet can act on a large set of inputs at one time. We could find a block, or set, of natural numbers (from column A) that produce the desired values in column B when the fill-down command is used.

Because we can use the fill-down command on values of *n* in the range $34 \leq n \leq 333$, the notion of variable is suddenly expanded from a single number to a subset of natural numbers. The function can then be used to find an "image set" of a subset of the domain containing more than one element.

The activity When a Variable Is Not Single allows us to begin thinking of a variable as more than a single number. The notion of variable is further expanded when we move from discrete to continuous functions: We can think about the domain and range of $f(n)$ extended to the set of real numbers, and then the graph of the function becomes continuous, as seen in figure 2.4. It is with continuous functions that we see the power of extending the concept of variable by considering a variable not as a single number but as the set of real numbers.

Fig. **2.4.**

The graph of $f(n) = 3n$, where n is a real number

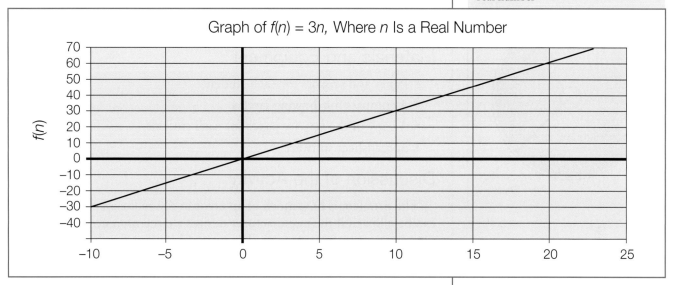

This power may also be demonstrated by mapping diagrams, which are used in the activity When a Variable Is Not Single to show what happens to subsets of the real numbers. An applied example involving tolerance is seen in the activity Tolerating a Bit.

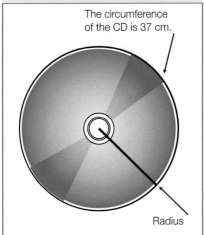

Fig. **2.5.**
A compact disc with a circumference of 37 centimeters

p. 56

The words *accuracy, tolerance,* and *precision* are frequently used interchangeably in the real world. In the original problem involving real-world company data, the words were interchanged.

Tolerating a Bit

Goals

- Examine a mapping diagram in a context
- Look at a function to see how changes in the domain affect the range
- Use class measurements to see how tolerance must be involved in measuring

Materials and Equipment

- Paper and scissors for each student
- A compass for drawing circles
- A copy of the activity pages for each student
- Spreadsheet software

Discussion of the Activity

The problem is how much of an error in the radius of compact discs can be allowed if the manufacturer desires a circumference of 37 centimeters, as seen in figure 2.5. The acceptable error for the manufacturer is sometimes described as *accuracy*, so we consider an *accuracy interval* about 37 centimeters. In order for the disc to be in the desired accuracy interval, the radius of the disk must be within a corresponding *tolerance interval*.

One of algebra's connections to measurement is reflected in the activity Tolerating a Bit. *Principles and Standards* (NCTM 2000) calls for students in grades 9–12 to be able to "apply appropriate techniques, tools, and formulas to determine measurements" by analyzing "precision, accuracy, and approximate error in measurement situations" (p. 320).

Extensions of Tolerance and Accuracy to Precalculus

To move from the arena of algebra 1 to algebraic notions of precalculus, it is possible to talk about an accuracy interval, or neighborhood, of size 0.002 about 9, as shown in figure 2.6. In the case of the function $f(x) = 3x$, the real-number images that fall in the given interval, or neighborhood, $(9 - 0.001, 9 + 0.001)$ are obtained when the preimages fall within a certain tolerance interval centered on 3. (Why?) How big can the interval around 3 be? This relationship can be modeled with mapping diagrams, as demonstrated earlier in this chapter, or with a geometric utility, as shown in the following activity, Tolerance and Accuracy.

Fig. **2.6.**
Accuracy interval, or neighborhood, about 9

Tolerance and Accuracy

Goals

- Use a graphic model to see how one tolerance affects another
- Examine inequalities in an application setting
- Examine an inequality solution geometrically

Materials and Equipment

- A copy of the activity sheets for each student
- Spreadsheet software
- A geometry utility

p. 59

Discussion of the Activity

This application of tolerance and accuracy uses specific numbers. The example can be generalized in an extension for upper-level algebra or precalculus.

Figure 2.7 summarizes the mathematical content of the activity in both English and algebraic symbolism. As long as x is in the given tolerance interval, then $f(x)$ is within the desired accuracy interval. Because in the real world repeating decimals are rarely handled, then the interval may become $2.9997 < 3 < 3.0003$, since $2.999\overline{6} < 2.9997 < 3 < 3.0003 < 3.000\overline{3}$. It is frequently easier mathematically to consider the last symmetrical interval about the desired value of $x = 3$ because it contains only terminating decimals. In terms of the application, to have an accuracy of 0.001 with the function $f(x) = 3x$, there has to be a tolerance of 0.0003 about x.

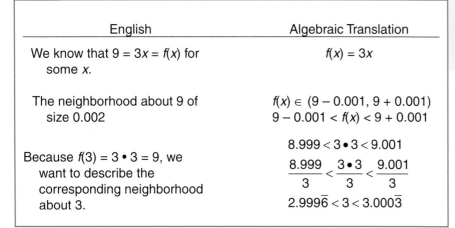

Fig. **2.7.**

An English and algebraic translation of Tolerance and Accuracy mathematics

If this example were generalized on a geometry utility, as in the activity Tolerance and Accuracy, students could discover what happens to every linear function of the form $f(x) = ax$, where a is a real number. As x changes, $f(x)$ changes proportionally.

Accuracy and tolerance can be discussed with absolute values. One definition of absolute value is $|x| = x$ if $x \geq 0$ and $-x$ if $x < 0$. The definition leads to the following inequality: $|x| < a$ implies $x < a$ if $x \geq 0$, and $x > -a$ if $x < 0$. Another way to write the sentence above is, $|x| < a$ implies $-a < x < a$. If the original problem is translated into inequality form with absolute values, then we have the question If $f(x) = 3x$, what is the value of x when $f(x)$ is within 0.001 of 9?

Chapter 2: Expanding the Notion of Variable

19

$$|f(x) - 9| < 0.001$$
$$|3x - 9| < 0.001$$
$$-0.001 < 3x - 9 < 0.001$$
$$9 - 0.001 < 3x < 9 + 0.001$$

The solution of the inequality can be continued as shown in the third entry in the chart in figure 2.7.

In Tolerance and Accuracy, the idea of what happens to a subset of real numbers in a business setting not only moves students to extend the notion of variable but develops a beginning understanding of the calculus of limits.

This chapter has explored the idea that a variable can be a set. After exposure to this idea, students no longer tie the concept to a single number, such as an input to a function or an unknown. This chapter opens the door to bigger ideas—for example, the distance function, which requires two inputs that are ordered pairs, or an area function, where the input is a triangle. Distance, area, and volume functions work in geometry just as functions do in algebra. The domains may be a set of ordered pairs, a set of two-dimensional figures, or a set of three-dimensional figures, respectively.

Variable takes on an expanded meaning in these mathematical contexts; students' understanding is not limited by the use and single context, as it was in the past; and a simple problem from the lower grades can be expanded to have even greater mathematical meaning.

Some Misconceptions about Variables

Rosnick, in "Some Misconceptions concerning the Concept of Variable" (NCTM 1999, p. 314), reported that more than 40 percent of the 152 sophomore and junior business majors taking a statistics class were unable to recognize P in the following multiple-choice questions.

> At this university, there are six times as many students as professors. This fact is represented by the equation $S = 6P$.
>
> A) In this equation, what does the letter P stand for?
>   i) Professors
>   ii) Professor
>   iii) Number of professors
>   iv) None of the above
>   v) More than one of the above (if so, indicate which ones)
>   vi) Don't know

A similar number of students could not identify S. Rosnick's results may indicate a lack of understanding of the relationship among the variables *students, professors,* the *number of students,* and the *number of professors,* but in any event, they indicate that students at the college level have a great deal of trouble understanding the meanings of variables. According to Rosnick, the students need to develop a more thorough understanding of the roles of variables and hence need either more or different opportunities to see variables in use. What has been proposed in this chapter is that teachers and students think about variables in different ways and see the idea of variable expanded to reflect its usage both in mathematics and in its applications.

Wagner (NCTM 1983/1999) says that teachers have an obligation to delimit the domain of a symbol and make very explicit what the domain is assumed to be if it is unspecified. In most cases, the unspecified domain is the set of real numbers or the "largest" subset of the real numbers that are usable in an example. However, misuse or nonspecification of domains has led to misunderstandings about functions that may be discrete but are assumed to be continuous. This type of misconception can lead to misinterpretations of data. An example from this book is the difference in the functions $f(x) = 3x$, where x is a natural number, and the function $f(x) = 3x$, where x is a real number. The range of the former describes a sequence of numbers, whereas that of the latter describes the set of all real numbers.

With the advent of computers, some equations like $n = n + 1$ that in algebra had no solutions now are used as an indexing tool or in a recursive format. This usage was seen in this chapter with the spreadsheet examples. Much like teachers' and students' notions and uses of variables, their ideas of equations and functions have to be expanded as uses of mathematics adapt to the real world.

NAVIGATIONS SERIES

GRADES 9–12

NAVIGATING through ALGEBRA

Chapter 3
Expanding the Notion of Function Representation

A recursive function is defined from an initial condition or conditions in such a way that later terms are defined in terms of earlier ones. For example, $f(1) = 4$; $f(n) = f(n – 1) + 3$, where n is a natural number, is a recursive function that produces the sequence 4, 7, 10, 13, …. An explicit function can be written in the form $y = f(x)$. For example, $y = x^2 + 1$ is an explicit function of x.

Processes can often be modeled discretely by sequences, as seen in chapter 2. Sequences are functions whose inputs are natural numbers. Sequences are useful in modeling situations in which the process can be described recursively. Such recursive descriptions, when coupled with the study of difference sequences (taking successive differences of a sequence of numbers) and web plots (graphs, seen later in this chapter, that follow the path of a function's orbit), can be very effective in studying the properties of the process and in generating explicit functions as models of the process. *Principles and Standards for School Mathematics* (National Council of Teachers of Mathematics [NCTM] 2000) calls for students to explore a variety of means of representing processes.

A search for a formula to model a situation uses the basic idea that arithmetic sequences have a common difference occurring when the first difference between successive terms is found. The pattern 3, 6, 9, …, studied earlier in this book, is a good example that students could consider. This pattern has a common difference of 3 between successive terms. As pointed out earlier, it leads to the concept of recursive functions. The "Add 3" pattern is a recursive pattern by its easy description: the "next" term is found by "adding 3" to the current term. Recursively, the formula could be described as

$$\text{NEXT} = \text{NOW} + 3$$

or as

$$\left\{ \begin{array}{l} y_1 = 3 \\ y_{n+1} = y_n + 3 \end{array} \right\}.$$

An explicit formula for the sequence is $f(n) = 3n$, where n is a natural number. As noted above, for this sequence, the first difference sequence contains only the number 3. Any such sequence that has a constant first difference sequence is linear and is of the form $f(n) = an + b$, where a and b are real numbers and n is a natural number.

For quadratic functions, two differences are required to reach a constant sequence difference. The notion of finite differences can be generalized for any polynomial function. An example is seen in figure 3.2. For geometric sequences, the notion of common differences does not work, as seen in figure 3.3. In the spreadsheet in figure 3.3, there is no apparent common difference, but looking diagonally down starting with column B, one sees the sequence 3, 9, 27, 81. If this sequence continues, it contains the powers of 3. The diagonal sequence directly underneath is 12, 36, 108, 324. These numbers are 4 times the powers of 3. The spreadsheet offers insight into what the sequence might be but gives no clear way to describe the sequence with a formula. A closer examination of column B shows that the terms might be described in a recursive way. For example, to define this geometric sequence, we could use the recursive formula

$$a_1 = 3,$$
$$a_n = 4a_{n-1}.$$

The recursive formula defines the sequence 3, 12, 48, ...

The function $f(n) = an + b$, where a and b are real numbers and n is a natural number, produces a constant-difference sequence (shown in the spreadsheet in fig. 3.1), each of whose terms is a, when n takes on successive natural numbers.

Fig. 3.1.
A constant-difference sequence as shown on a spreadsheet

	A	B	C
1	n	f(n)	Difference
2	1	a + b	
3	2	2a + b	(2a+b)−(a+b)=a
4	3	3a + b	(3a+b)−(2a+b)=a
5	4	4a + b	(4a+b)−(3a+b)=a

Fig. 3.2.
Successive differences seen on a spreadsheet

	A	B	C
1	n	Term	Difference in Successive Terms
2	1	3	
3	2	6	3
4	3	9	3
5	4	12	3
6	5	15	3
7	6	18	3

(a) Difference for a linear function

	A	B	C	D
1	n	Term	First Difference in Successive Terms	Second Difference in Successive Terms
2	1	3		
3	2	6	3	
4	3	10	4	1
5	4	15	5	1
6	5	21	6	1
7	6	28	7	1

(b) Difference for a quadratic sequence

The Fibonacci sequence is typically shown as 1, 1, 2, 3, 5, 8, 13, ...

Difference equations and a computer algebra system (CAS) or spreadsheet can be used to develop formulas for polynomial patterns or some other patterns, such as the Fibonacci sequence. Using a sequence

Fig. **3.3.**
A spreadsheet showing differences for a geometric sequence

	A	B	C	D	E
1	n	Term	First Difference in Successive Terms	Second Difference in Successive Terms	Third Difference i
2	1	3			
3	2	12	9		
4	3	48	36	27	
5	4	192	144	108	81
6	5	768	576	432	324
7	6	3072	2304	1728	1296
8	7	12288	9216	6912	5184
9	8	49152	36864	27648	20736
10	9	196608	147456	110592	82944

of differences to obtain a formula to represent a set of data is seen again in chapter 5. Good models for exploring difference equations and sequences of differences are found in *Mission Mathematics, Grades 9–12* (House 1997, pp. 21–26).

Recursive formulas and difference sequences are examples of *iterative change*. Iterative change has many appealing possibilities for algebraic investigation. The following examples of recursive processes help to illustrate the benefits of this approach.

The first example in the activity Would You Work for Me? was adapted from an operations-research problem from decades ago.

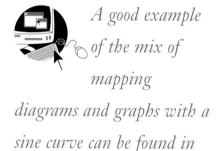

A good example of the mix of mapping diagrams and graphs with a sine curve can be found in Brieske (1980, p. 275).

Finite Differences, a spreadsheet applet on the CD-ROM, allows users to explore difference sequences.

Would You Work for Me?

Goals

- Use recursive or iterative forms to represent relationships
- Approximate and interpret rates of change from numerical data
- Draw reasonable conclusions about a situation being modeled

p. 63

Materials and Equipment

- A copy of the activity pages for each student
- Spreadsheet software or a graphing calculator, if needed

Activity

Would you work for me or for my sister with the following salary schemes? We will pay you $1 for your first day's work and $0.50 for the second day's work. Each day after the second, your salary will be computed as follows:

$$\text{Tomorrow's salary} = \left(2\tfrac{1}{2}\right)(\text{Today's salary}) - \text{Yesterday's salary}$$

Furthermore, we don't like to bother with pennies but disagree on how they should be dealt with in computing your salary. As the glass-is-half-full optimist, I always round up to the nearest dime. My sister, however, as the glass-is-half-empty person, always rounds down to the nearest dime. Would you work for either of us? Why or why not?

Discussion

This problem can be solved with technology; technology handles recursive relationships well, and graphing each salary scheme is interesting. The equations are simple and based on a geometric relationship.

The second problem has been called The Devil and Daniel Webster. It was suggested by Boyd Henry of the College of Idaho.

The Devil and Daniel Webster

Goals

- Use recursive or iterative forms to represent relationships
- Approximate and interpret rates of change from numerical data
- Draw reasonable conclusions about a situation being modeled

Materials and Equipment

- A copy of the activity page for each student
- Spreadsheet software or graphing calculator, if needed

Activity

The devil made a proposition to Daniel Webster. The devil proposed paying Daniel for services in the following way:

On the first day, I will pay you $1000 early in the morning. At the end of the day, you must pay me a commission of $100. We will determine your next day's salary and my commission according to the following rule: I will double what you have at the end of the day, but you must double the amount that you pay me. Will you work for me for a month?

Discussion

Again, recursion is the order of the day. Graphing results is challenging. Understanding the type of change involved is also challenging. Variations on this problem might include having the devil ask for a cut of 80 percent of the salary each day instead of the doubling scheme. Should Daniel do the work for three weeks? For a month? At what rate is it most advantageous for both Daniel and the devil? Should Daniel work if there were a 1 percent cut for the Devil? For advanced students, the question of developing a closed formula might be posed.

Principles and Standards (NCTM 2000) emphasizes that recursive formulas are used to solve many problems and that students often have a natural understanding of recursively defined functions. The final example in this chapter is developed from a classic estimation problem. Consider an iterative method for finding the square root of a positive real number. To find $\sqrt{2}$, the "Babylonian method" (Wagner 1998) tells one to start with a guess that is larger than $\sqrt{2}$; for example, guess 2. The "next" guess is generated by the recursive algebraic process

$$\text{NEXT} = \frac{\text{CURRENT} + \dfrac{2}{\text{CURRENT}}}{2}.$$

This iterative process can be executed on a calculator and iterations achieved by simply pressing the enter key. The activity Iterating to Find the Square Root of 2 outlines this process for students.

p. 66

The Babylonian method refers to a clay tablet that gives some of the earliest results of explorations of $\sqrt{2}$. "The tablet consists simply of the drawing of a square with its two diagonals. Three numbers in sexagesimal notation are given: 30 along one of the sides and the two numbers 1; 24,51,10 and 42; 25,35 along one of the diagonals. [*Ed. note:* 1; 24,51,10 would be 1 + 24/60 + 51/3600 + 10/216000.] It is easily verified that 1; 24,51,10 is an excellent approximation for $\sqrt{2}$, while the value 42; 25,35 is the product of the side 30 with the given approximation. From this it may be deduced that the early Babylonians possessed arithmetical techniques sufficient for obtaining a good approximation to $\sqrt{2}$..." (NCTM 1969, p. 100).

Iterating to Find the Square Root of 2

Goals
- Investigate how technology makes an old algorithm easy to use
- Use different modes to see how effective an iterative algorithm is

Materials and Equipment
- A computer algebra system, an appropriate calculator, or spreadsheet software
- A copy of the activity pages for each student

p. 69

Discussion of the Activity

Consider extension questions 3 and 4 on "Iterating to Find the Square Root of 2":

3. How many iterations would be required to get accuracy to one hundred decimal places in the estimate of $\sqrt{2}$?

4. If the process converges to $\sqrt{2}$, how "fast" does it converge? That is, how does the error change from iteration to iteration?

To address these questions, students at advanced levels need to explore number and operation in depth. For example, after eight iterations, the exact result of the process is a formidable fraction. Students can use a computer algebra system to get the first 300 decimal places in the expansion of this fraction. To see how close this approximation is, we square the approximation, N, and compare the result with 2. If the approximation is squared, the result is 2.0..., with 155 zeros before the next nonzero digit. Counting the 155 zeros after the 2, we can say that $|N^2 - 2| \leq 10^{-155}$. Solving this inequality, we get $|(N - \sqrt{2})(N + \sqrt{2})| \leq 10^{-155}$. Therefore,

$$\left|N - \sqrt{2}\right| \leq \frac{10^{-155}}{N + \sqrt{2}} \leq 10^{-155}.$$

See Devany (1991) for further insights into iteration.

Thus, the approximation to $\sqrt{2}$ is accurate to at least 154 decimal places. The technology enables advanced students to apply not only their understanding of number, place value, and operation but also their knowledge of factoring, inequalities, absolute value, and mathematical reasoning. See also the discussion of a web plot in figure 3.4.

This method of expanding a fraction to 300 decimal places can be used by students to study the repeating-decimal phenomenon of rational numbers and to explore the irrationality of $\sqrt{2}$. This process fits with the recommendation in *Principles and Standards* (NCTM 2000) for high school students to explore system properties of numbers. By applying the rational-root theorem found in many algebra 2 textbooks, students discover that the equation $x^2 - 2 = 0$ has no rational roots. Therefore, $\sqrt{2}$ must be irrational, since it is a root of that equation. CAS can be used by students in the development of a proof of the rational-root

A web plot models the path created by the orbit of a function. The orbit of a function $f(x)$ is created by taking an initial value of x, such as x_0 (or x_1), and creating the sequence $x_0, f(x_0), f(f(x_0)), \ldots$ (or $x_1, f(x_1), f(f(x_1)), \ldots$). For example, in the drawing, choose (x_0, x_0) as a point on $y = x$ to start. The tip of the uppermost black arrow is $(x_0, f(x_0))$. Going directly left leads to $(f(x_0), f(x_0))$ on the line $y = x$. Repeating this process produces the web plot "from the top." In this case, regardless of where you start, the web plots "from the top," starting at (x_0, x_0), and "from the bottom," starting at (x_1, x_1), converge at $(1, 1.00)$.

Fig. **3.4.**

The web plot modeling the orbit of $f(x) = \sqrt{x}$

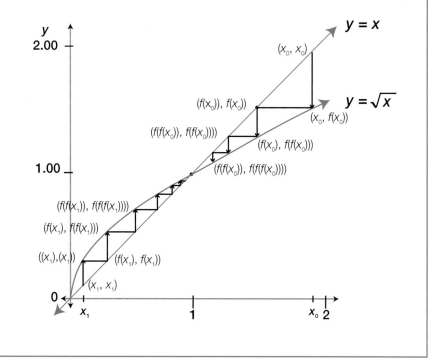

theorem, the factor theorem, the remainder theorem, and other results for polynomials.

To address question 4, students can use the technique above to count the number of accurate decimal places in their answers after one, two, three, …, eight iterations and make a table. The discussion of how to measure accuracy is important, but good estimates of the number of accurate decimal places are sufficient. Students can look at various regression options using the values in their tables. An exponential regression fits the data well, indicating that the number of digits of accuracy approximately doubles with each iteration. Translating this result into a statement about the error at each step, students can generate a recursive formula for the error: $E(n) \cong E(n-1)^2$.

With technology, one might test the pattern
$$E(n) \cong E(n-1)^2$$
using algebraic processes: if $g = $ *guess* at stage $n-1$, with $g > \sqrt{2}$, and
$$E(n-1) = g - \sqrt{2}$$
is the error at that stage and
$$E(n) = \frac{g + \frac{2}{g}}{2} - \sqrt{2}$$
is the error at the next stage, then the ratio
$$\frac{E(n)}{E(n-1)}$$
is as shown. Thus,
$$\frac{E(n)}{E(n-1)} = \frac{E(n-1)}{2g},$$
or
$$E(n) = \frac{E(n-1)^2}{2g}.$$

Assuming g is greater than 1, we get $E(n) < E(n-1)^2$, thus confirming the pattern found in the students' tables.

Navigations Series

Grades 9–12

Navigating through Algebra

Chapter 4

Expanding Students' Understanding of Algebraic Equivalence or Identity

Principles and Standards for School Mathematics (National Council of Teachers of Mathematics [NCTM] 2000) calls for students to expand their understanding of the processes used in algebraic manipulations. Such an expansion leads to the multimodal view of mathematical (or algebraic) literacy. Teachers must constantly pay attention to the multimodal nature of mathematical understanding. The learning of algebraic processes is enhanced for all students when the following guidelines are adhered to:

1. The student understands the overall goal of the algebraic process and knows how to predict or estimate the outcome.
2. The student understands how to carry out an algebraic process and knows alternative methods and *representations* of the process.
3. The student understands and can *communicate* to others why the process is effective and leads to valid results.
4. The student understands how to evaluate the results of an algebraic process by invoking *connections* with a context or with other mathematics the student knows.
5. The student understands and uses mathematical *reasoning* to assess the relative efficiency and accuracy of an algebraic process compared with alternative methods that might have been used.
6. The student understands why an algebraic process empowers her or him as a mathematical *problem solver*.

These six guidelines are essential elements in the teaching and learning of algebra. If students are taught and asked to demonstrate a knowledge of only how to carry out an algebraic process, then their understanding of the process will usually be inadequate. When these guidelines are incorporated into algebraic investigations of various contexts, conceptual versus procedural understanding becomes a false dichotomy and greater fluency in the use of variable, operation, and relation can be achieved. Technology supports the development of multiple representations of algebraic processes as well as aids in achieving the fluency and proficiency goals set out in *Principles and Standards* (NCTM 2000, pp. 35, 39, 294, 301).

As with most change in education, the use of technology in mathematics has been controversial. By including the Technology Principle, *Principles and Standards* (NCTM 2000) accepted the research conclusion that "students can learn more mathematics more deeply with the appropriate use of technology" (p. 25). However, *Principles and Standards* also states that "technology should not be used as a replacement for basic understandings and intuitions; rather, it can and should be used to foster those understandings and intuitions" (p. 25). What follows in this chapter is a mix of examples and activities that can be done by hand or using technology. The authors strongly believe that the two should be done in tandem to achieve real understanding of the algebra involved.

Spreadsheets, mapping diagrams, and geometry utilities are aids in achieving an expanded notion of function. As we consider the process of working with operations on functions and of building on prealgebra experiences, we consider how operations and an understanding of them expand for a student. The language of algebra generalizes what was beginning to be learned with arithmetic processes in the lower grades. As pointed out by Demana and Leitzel (1988), "students can understand basic concepts of algebra when they are [or have been] introduced through numerical computation and problem solving before they are encountered in more formal courses in algebra" (p. 61). This chapter is written under the assumption that in lower grades, students have been introduced to algebra in the way suggested by Demana and Leitzel.

As an example of equation solutions that students might have encountered in the middle grades, consider the solution of the following equation:

$$5(3 - x) + 4 = 2x - 9 \qquad (1)$$
$$15 - 5x + 4 = 2x - 9$$
$$-5x + 19 - 2x = -9 \qquad (2)$$
$$-7x = -28 \qquad (3)$$
$$x = 4 \qquad (4)$$

We expect students to be able to solve equations such as those above with pencil and paper. A computer algebra system will solve the equation immediately, but it imparts little understanding of the process, and solving the equation in a rote manner requires no understanding. *Principles and Standards* (NCTM 2000, p. 296) states that all students should "represent and analyze mathematical situations and structures using algebraic symbols" with an understanding of "the meaning of equivalent forms of expressions, equations, inequalities, and relations."

Different representations of the algebraic notions involved in solving equation 1, above, may help students develop this understanding. In the activity Representing the Solution Process by Graphing, either a graphing calculator or a computer graphing package is an appropriate tool to use to investigate the process.

Representing the Solution Process by Graphing

Goals
- Graphically represent the solution to a simple equation
- Explore the meaning of equivalent equations

Materials and Equipment
- A copy of the activity pages for each student
- A graphing calculator or a computer graphing package

Discussion of the Activity

Representing the Solution Process by Graphing leads students through the thought process behind solving an equation graphically in two different ways. The activity assumes that while solving an equation, students are working with systems of equations. The students need to understand that each successive equation in the solution should be equivalent to the preceding one. (If two equations, or systems of equations, are *equivalent*, then they have the same solution.) Suppose each side of one of the equations in the solution on page 32 is thought of as an equation (or function). The two sides then form a system of equations. For example, the first equation in the solution is

$$5(3 - x) + 4 = 2x - 9.$$

From this one equation, we obtain the following system:

$$\begin{cases} y_1 = 5(3-x) + 4 \\ y_2 = 2x - 9 \end{cases}$$

Each single equation in the solution similarly generates a system of equations:

(2) yields y_3 and y_4
(3) yields y_5 and y_6
(4) yields y_7 and y_8

In the complete solution shown, the systems should be equivalent if the answer truly is the solution to the original equation. In fact in the activity, students should realize that what the systems have in common is that the solution to the system has the same x-coordinate in each case. That x-coordinate (4) is the solution to the original equation. Note that for the solution we are not interested in the value of the y-coordinate. We can examine each graph separately, as in figure 4.1a, or look at all the systems on the same graph, as seen in figure 4.1b.

The graphs are dependent on the steps in the solution. Different solutions may produce different graphs, but the value of x should be the same in all the solutions.

An extension in the activity considers a situation in which the final answer is not a solution to the original equation and asks students to explain why.

p.72

Representing the Solution Process by Graphing is designed for limited use in helping students see graphically what happens when an equation is solved. It is not suggested that this activity be done on more than two equations.

In the extension, the student is asked to solve

$$\frac{x^2 - 9}{x - 3} = 0.$$

Working by hand following routine procedures, a student may find that $x = +3$ or -3. But $+3$ cannot be a solution because we would have division by 0 in the denominator of the original expression on the left side of the equation. This is seen in the systems approach using technology.

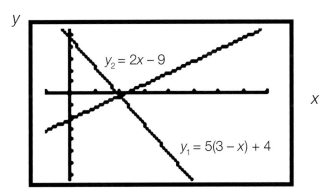

(a) A graph of the system of equations generated by the original equation

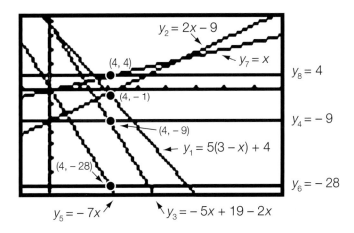

(b) A graph of the systems of equations generated by the steps in the solution

Fig. **4.1.**

Graphs seen in the solution of
$5(3 - x) + 4 = 2x - 9$

y_1 and y_2 intersect at $(4, -1)$
y_3 and y_4 intersect at $(4, -9)$
y_5 and y_6 intersect at $(4, -28)$
y_7 and y_8 intersect at $(4, 4)$

The idea of using a series of simultaneous equations to examine the solution to a simple equation gives students a new understanding of what equivalent equations are; that is, if two or more equations in a system are equivalent, then their solutions are the same. The same idea can be used to examine the solution of an inequality. Consider this inequality solution:

$$5(3 - x) + 4 \leq 2x - 9$$
$$15 - 5x + 4 \leq 2x - 9$$
$$19 - 5x \leq 2x - 9$$
$$19 + 9 \leq 2x + 5x$$
$$28 \leq 7x$$
$$4 \leq x$$

If we consider pairs of simultaneous equations from the inequality solution above, as we did in Representing the Solution Process by Graphing for the equation solution, then the graphical representation will let us estimate the approximate solution of a series of inequalities that are equivalent. The graph in figure 4.2 shows $y_1 \leq y_2$ when $x \geq 4$. Thus the solution to the original inequality is the set of all values of x such that $x \geq 4$. Note that we are not interested in the values of the y-coordinate.

Fig. **4.2.**

The solution to $y_1 \leq y_2$

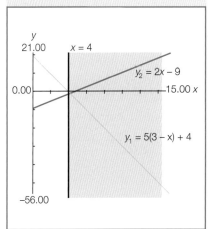

Chapter 4: Expanding Understanding of Algebraic Equivalence or Identity

The techniques illustrated for equations or inequalities consider equivalent systems of equations in their solution of simple equations. The same techniques can be applied to the solution of an equation such as the following:

$$\sqrt{5x^2 - 36} = x$$
$$5x^2 - 36 = x^2$$
$$4x^2 = 36$$
$$x^2 = 9$$
$$x = 3$$

or

$$x = -3$$

Looking at the graphs of the systems of equations in the solution gives us the graphs in figure 4.3. (Note that in figure 4.3, the software was allowed to set the scales on the axes to fit the computer screen on which each graph was drawn. Students or teachers may set their own scales

Fig. 4.3.

In the typical algebraic solution to $\sqrt{5x^2 - 36} = x$, something strange happens.

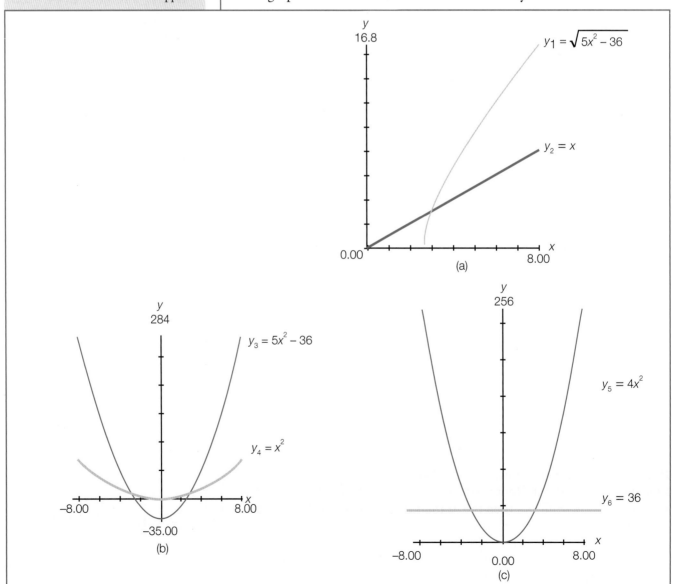

when using graphing software.) From the graphs, we can see that when both sides of the equation were squared in an attempt to find the algebraic solution, an additional solution was introduced that does not satisfy the original equation. The solution to the original equation is $x = 3$, whereas $x = -3$ is not a solution to the original equation. This use of graphs should immediately convince students that the first two equations in the algebraic solution are not equivalent. Using a graphical representation in concert with an algebraic solution encourages sense making as well as prompts a discussion of extraneous roots.

One of the expectations for all students in grades 9–12 listed in *Principles and Standards* (NCTM 2000, p. 296) is that they "understand and compare the properties of classes of functions, including exponential, polynomial, rational, logarithmic, and periodic functions." Technology offers tools with which to think about families of functions, as demonstraed in the next activity, Families of Functions.

Systems of Equations is an applet on the CD-ROM that assists users in solving problems graphically as well as algebraically.

Families of Functions

Goals

- Recognize properties of families of functions
- Apply properties of functions

Materials and Equipment

- A copy of the activity pages for each student
- Spreadsheet software or a graphing calculator

Discussion of the Activity

In this activity, the problems involving median salary data for men and women over a seven-year period may provide an application setting for examining families of linear functions. The data in table 4.1 were taken from the U.S. Census (www.census.gov/income/p13.txt). If students find a linear equation to describe the men's data, for example, and the women's data line seems to be shifted below the men's line, as seen in figure 4.4, determining what can be done to the one equation to find the other leads to an examination of the effect of b in the general linear equation $y = mx + b$.

Table 4.1
Median Salaries for Workers Twenty-five Years or Older according to Gender

Year	Median Salary: Women	Median Salary: Men
1991	$11,580	$23,686
1992	$11,922	$23,894
1993	$12,234	$24,605
1994	$12,766	$25,465
1995	$13,821	$26,346
1996	$14,682	$27,248
1997	$15,573	$28,919

Part 3 of the activity may evoke a discussion of the interconnection of statistics and algebra. At this stage, the teacher may want to discuss the use of residuals for finding a "line of best fit" or a "curve of best fit." The teacher may choose to use a "spaghetti fit" line, in which a piece of uncooked spaghetti is used to best approximate the trend of the data.

Spreadsheets are very helpful in deciding for what domains operations—such as addition, subtraction, multiplication, division, or composition—on functions can be computed. If spreadsheets are used to find a solution, then it becomes clear when some computations can be done or cannot be done. Consider, for example, $f(x) = x^2$ and $g(x) = x - 1$. Adding, subtracting, and multiplying these two functions either on mapping diagrams or on a spreadsheet are relatively easy, as seen in the activity Operating on Functions.

This activity is based on a idea from *Contemporary Mathematics in Context: A Unified Approach* (Coxford et al. 1997, pp. 203–4).

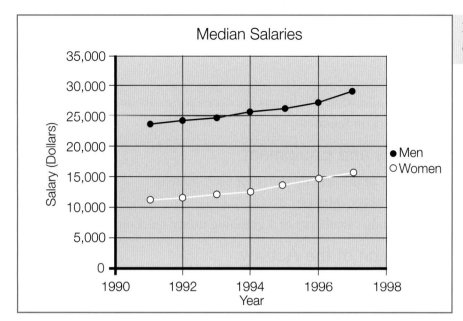

Fig. **4.4.**
Graphs of median-salary data

Chapter 4: Expanding Understanding of Algebraic Equivalence or Identity

Operating on Functions

Goals

- Recognize what conditions are necessary to divide functions
- Operate on functions

Materials and Equipment

p. 80

- A copy of the activity pages for each student
- Spreadsheet software
- Graphing calculators or a graphing utility on computers

Discussion of the Activity

Evaluating either quotient $f(x)/g(x)$ or $g(x)/f(x)$ with a spreadsheet yields undefined results when division by 0 is encountered. Knowing that the denominator of a fraction cannot be zero and that division is defined only where the domains of the numerators and denominators intersect becomes more important. If a student tries to divide $f(x) = x^2$ by $g(x) = x - 1$, the technology should signal that something is awry.

NAVIGATIONS SERIES

GRADES 9–12

NAVIGATING through ALGEBRA

Chapter 5
Expanding Students' Understanding of Change

According to Hahn (1998, pp. 76–78), in 1604 and 1606, Galileo Galilei conducted a series of experiments involving rolling a metal ball down an inclined plane. Galileo's theory was that all bodies fall with the same acceleration regardless of their weight as long as they are heavy enough and shaped in such a way that air resistance can be ignored. Galileo discovered that the acceleration of the ball on the inclined plane was constant and that the trajectory of a ball, after rolling off the end of an inclined plane situated above the ground, was parabolic.

Students need to study the change attributes of different families of functions and processes. They can do so by using algebraic processes to study changes in the world around them and, reciprocally, by using the change attributes of processes to understand algebraic processes or functions better. By modeling real-world phenomena using curve fitting and simulation, students learn to identify and exploit the change attributes of mathematical and real-world processes.

Advances in classroom technology have made the study of mathematical change readily accessible to students. Motion detectors that measure the distance of an object from the device many times in a single second allow students to gather distance-time data on moving objects. In order to analyze the graph of a set of distance-time data, the students must be able to interpret the rates of change in distance, velocity, and acceleration that are observable in a graph. For example, students might imitate the Galileo experiments on inclined planes.

The set of data in table 5.1 was gathered by rolling a ball down an inclined plane. Students can study the motion of the ball rolling down the plane by considering the difference sequences of the distance data in table 5.1, as shown in table 5.2.

Galileo predicted that the velocity of the moving object would increase at a constant rate, or in other words, the acceleration of such a moving object is constant. Galileo's prediction implies that the first difference sequence increases at a constant rate. Students may test Galileo's hypothesis that acceleration is constant against the data in table 5.2. It is clear by looking at the sequence of second differences that the

The average velocity in each interval can be found from table 5.2 by dividing the terms of the first difference sequence by the change in time (0.04 seconds).

Table 5.1
Time and Distance for Ball Rolling down a Plane

Time (Seconds)	Distance (Meters)
0	0
0.04	0.0318
0.08	0.067
0.12	0.1043
0.16	0.1448
0.2	0.1877
0.24	0.236
0.28	0.2854
0.32	0.337
0.36	0.3907
0.4	0.4489
0.44	0.5093
0.48	0.5729
0.52	0.6454
0.56	0.7112
0.6	0.7826
0.64	0.8572
0.68	0.9341
0.72	1.0131
0.76	1.0965

The Finite Differences applet on the CD-ROM can extend explorations of difference sequences.

Table 5.2
Time and Distance for a Ball Rolling down a Plane with Two Difference Sequences

Time (Seconds)	Distance (Meters)	First Difference Sequence of Distances	Second Difference Sequence of Distances
0	0		
0.04	0.0318	0.0318	
0.08	0.0667	0.0349	0.0031
0.12	0.1043	0.0376	0.0027
0.16	0.1448	0.0405	0.0029
0.2	0.1877	0.0429	0.0024
0.24	0.236	0.0483	0.0054
0.28	0.2854	0.0494	0.0011
0.32	0.337	0.0516	0.0022
0.36	0.3907	0.0537	0.0021
0.4	0.4489	0.0582	0.0045
0.44	0.5093	0.0604	0.0022
0.48	0.5729	0.0636	0.0032
0.52	0.6454	0.0725	0.0089
0.56	0.7112	0.0658	-0.0067
0.6	0.7826	0.0714	0.0056
0.64	0.8572	0.0746	0.0032
0.68	0.9341	0.0769	0.0023
0.72	1.0131	0.079	0.0021
0.76	1.0965	0.0834	0.0044

differences are not constant. However, if Galileo's claim is correct, then the fluctuations might be attributed to various errors, including measurement errors. If we allow for measurement errors, we could reasonably claim that a constant difference is the average of the terms of the second difference sequence. Thus, the constant difference could be approximated as 0.0029.

Any sequence of values in which the second difference sequence produces a constant difference can be represented by a quadratic formula of the form $f(t) = at^2 + bt + c$. In this case, Galileo's prediction implies that the distances are represented by a quadratic. Table 5.3 shows differences arising from the quadratic $f(t)$ where t takes on the given values. The third column shows the first differences of the quadratic expression, and the fourth column shows the second differences.

Table 5.3
Differences Arising from a Quadratic Function

Time	The Form of a Quadratic Representing Distance	First Difference	Second Difference
0	$a(0)^2 + b(0) + c = c$		
0.04	$a(0.04)^2 + b(0.04) + c$ $= 0.0016a + 0.04b + c$	$0.0016a + 0.04b$	
0.08	$a(0.08)^2 + b(0.08) + c$ $= 0.0064a + 0.08b + c$	$0.0048a + 0.04b$	$0.0032a$
0.12	$a(0.12)^2 + b(0.12) + c$ $= 0.0144a + 0.12b + c$	$0.0080a + 0.04b$	$0.0032a$

If we set the second difference in table 5.3 equal to 0.0029, then we have

$$0.0032a = 0.0029,$$
$$a = 0.90625.$$

Now

$$0.0016a + 0.04b = 0.0318,$$

but $a = 0.90625$, so

$$(0.0016)(0.90625) + 0.04b = 0.0318,$$
$$b = 0.7588.$$

Further, $c = 0$ from the first difference. Hence, the following quadratic emerges as a candidate for a model under Galileo's prediction of constant acceleration: $f(t) = 0.90625t^2 + 0.75875t$, where $f(t)$ is the distance function for the data. The graph in figure 5.1 confirms that the model is indeed a good fit.

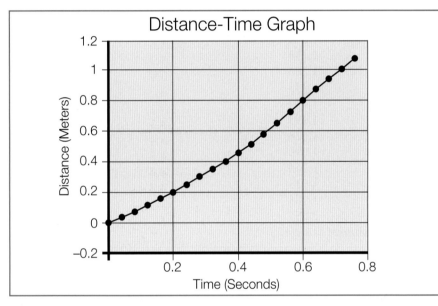

Fig. **5.1.**

A distance-time graph with a regression equation

The development of calculus was fueled by the needs of scientists to come to grips with the notions of instantaneous velocity and acceleration pointed to in the work of Galileo and others.

The preceding example illustrates that students can use first- and second-order difference sequences to study changes in a sequence of data. Algebraic approaches using difference sequences and equations, assisted by technology, open up whole new avenues for studying concepts of change.

Students can become fluent with a variety of models and their change patterns, as illustrated in the following example, which is adapted from a problem found in *Calculus Concepts: An Informal Approach to the Mathematics of Change*, by LaTorre et al. (1995, p. 189).

Table 5.4 gives rates of deaths due to lung cancer (deaths per 100,000 males in the United States) every ten years from 1930 to 1990. Students can discuss why linear, quadratic, cubic, or exponential models are or are not appropriate for this data set. Of the models listed, they should choose the one that is the best for these data and then fit the model chosen to the data, give the equation used, and use the model to predict the death rate in 2000. They could discuss whether the model is appropriate for future predictions.

Not only can change be studied from the quantitative perspective of difference sequences, it can also be simulated by the use of parametric equations with the help of graphing technologies. The following

Table 5.4
Rates of Deaths Due to Lung Cancer

Year	Death Rate of Males per 100 000
1930	5
1940	11
1950	21
1960	39
1970	59
1980	66
1990	67

example illustrates some simple activities that require students to focus on the change attributes of three situations. The first situation involves asking students to model the motion of a car that is traveling at a constant speed. With a motion detector measuring their distance from a starting point (a wall), students gather distance-time data by moving away from the wall. Sample data collected by a student are given in table 5.5.

Table 5.5
Distance-Time Data Gathered by a Motion Detector

Time (Seconds)	Distance (Meters)	Time (Seconds)	Distance (Meters)
0	0.491	2.5	2.12
0.2	0.591	2.7	2.259
0.3	0.648	2.8	2.328
0.4	0.708	2.9	2.399
0.5	0.768	3.1	2.533
0.6	0.829	3.2	2.601
0.8	0.959	3.3	2.672
0.9	1.026	3.5	2.819
1	1.094	3.6	2.886
1.1	1.158	3.7	2.957
1.3	1.286	3.9	3.101
1.4	1.353	4	3.177
1.5	1.424	4.1	3.25
1.6	1.493	4.3	3.392
1.8	1.628	4.4	3.456
1.9	1.694	4.5	3.516
2	1.764	4.7	3.641
2.2	1.913	4.8	3.705
2.3	1.985	5	3.826
2.4	2.055	5.1	3.882
(Continued at right)			

The data in table 5.5 are shown in the graph in figure 5.2. Students' examination of the graph of the data they gathered can lead to a discussion of distance-time graphs and the relation of slope to velocity. Instead of using linear regression to fit a line to these data, students can choose two representative points on the graph and find the line passing through them. Such curve-fitting problems easily lend themselves to discussions of families of functions. In this case, students might discuss the influence of the parameters m and b in the linear equation $y = mx + b$. The equation of the regression line for the data in table 5.5 is $y = 0.685x + 0.419$, and its graph closely fits the scatterplot in figure 5.2, as seen in figure 5.3.

Students can simulate the motion from the distance-time graph using a set of parametric equations. To simulate the motion along a straight line using a horizontal path through the middle of a calculator screen, students fix the value of y as a constant and form parametric equations using the regression-line equation for the data in table 5.5 to form $x(t)$, as follows:

$$x(t) = 0.685t + 0.419$$
$$y(t) = 5$$

Using the parametric equations above for the path of the object, students can then generate a dynamic graph with the cursor moving across the screen in a straight line with a velocity determined by the slope of the function $x(t)$ at the point. The cursor's position at equal time intervals is marked by a dot on the graph in figure 5.4.

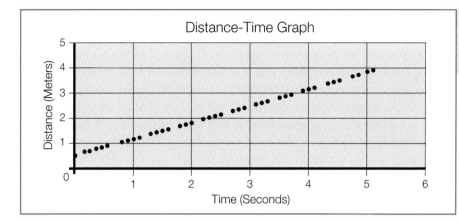

Fig. **5.2.**

A distance-time graph of data collected with a motion detector

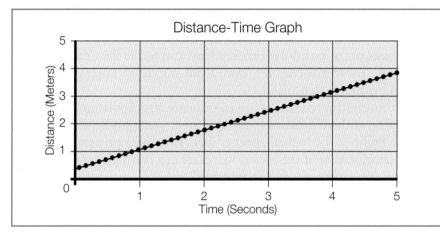

Fig. **5.3.**

The distance-time graph with a regression line

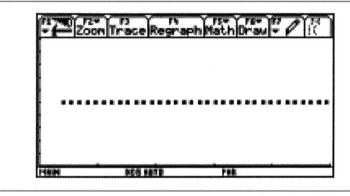

Fig. **5.4.**

A simulation of the motion in the distance-time graph

After this exploration, students could be challenged to produce the parametric equations that represent the motion of a car moving at 35 miles per hour. Letting time t be measured in hours, the parametric equations that they obtain might be the following:

$$x(t) = 35t$$
$$y(t) = 5$$

With time expressed in seconds and distance in feet, students have to convert miles per hour to feet per second to get parametric equations such as the following:

$$x(t) = 51.33t$$
$$y(t) = 5$$

In a similar fashion, students might be challenged to study and simulate the motion of a car along a straight line starting at a full stop and accelerating constantly to 55 miles per hour. From the study of difference sequences for polynomial equations suggested in the Galileo experiment, students can conclude that a model with constant acceleration would be quadratic, and they could experiment with this model in mind. Figure 5.5a shows a scatterplot produced by a student attempting to move a motion detector away from a wall at a constant acceleration. Using family-of-function ideas and curve-fitting techniques such as quadratic regression, students can produce models such as $d(t) = 0.171t^2 + 0.204t + 0.413$ for the distance-time equation. This model closely fits the data, as seen in figure 5.5b. To simulate this motion using parametric equations, students would develop the following equations:

$$x(t) = 0.171t^2 + 0.204t + 0.413$$
$$y(t) = 5$$

On the basis of such explorations, students can tackle the challenge of simulating a car that accelerates from a full stop to 55 miles per hour and then maintains that speed.

Other interesting variations are to simulate a rocket launch in a vertical direction with an initial velocity specified. Once students have simulated the flight of the rocket, another scenario can be presented: Specify that the acceleration of the rocket increases at a constant rate while the fuel burns; that when the fuel is expended, the rocket glides to its maximum height and then begins its descent; and that when the rocket's parachute opens, it glides to the ground at a constant rate of speed. Another interesting variation is to apply the parametric way of visualizing change to situations in which the processes are defined iteratively, such as investment plans and the growth of money in a savings account under various compoundings of interest.

All problems in this chapter deal with analyzing algebraic processes that quantify the change attributes of the process. They are designed to help students understand how vastly different rates can be and how patterns can change with minor adjustments.

Fig. **5.5.**

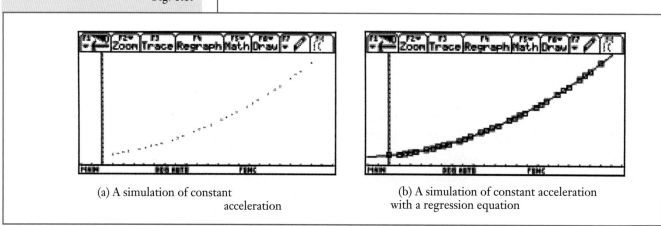

(a) A simulation of constant acceleration

(b) A simulation of constant acceleration with a regression equation

Navigating through Algebra

Grades 9–12

Chapter 6
Conclusion

This book has attempted to demonstrate how algebra is used to describe mathematical and real-world processes and to study their properties. We focused on the basic elements of algebra found in the concepts of variable, operation, function, and relation. As one considers the diverse array of possibilities envisioned in *Principles and Standards for School Mathematics* (NCTM 2000), an underlying theme is the notion that regardless of the topic or algebraic focus of the student's activity, the goals for understanding must be multimodal.

To achieve the goals of multimodal understanding described in the book, assessment must be embedded in all mathematical activities. When students are exploring real-world or mathematical processes, which often take significant amounts of class time, the essential elements of understanding need to be assessed before the end of the activity to ensure successful learning. Teachers must therefore become adept at the art of questioning and be prepared to take appropriate measures when their questions reveal a lack of understanding at crucial stages in the activity.

With simple and general rubrics, teachers can help students develop mature algebraic-reasoning skills. The examples in this book offer a sample of the diversity of contexts in which algebraic reasoning can be developed. The examples also show how teachers can explore more deeply the basic algebraic elements of variable, operation, function, and relation.

NAVIGATIONS SERIES

GRADES 9–12

NAVIGATING through ALGEBRA

Appendix
Blackline Masters and Solutions

What's the Area?

Name _____

A subdivision is being placed on a piece of land 1000 m by 1500 m. A boulevard of trees and an access road of uniform width form the border of the subdivision, as shown. The area of the inner rectangle of houses and parks is to be at least 1.35 million m² to accommodate the planned homes and parks. What is the largest width that can be set aside inside the perimeter for the border composed of the boulevard and road?

Border (Boulevard and Access Road)

Area of Houses and Parks

1. Make guesses about the width of the border in the problem and keep a record of the width and the resulting area for the houses and parks. (Your recordings may be done on a spreadsheet or by hand.)

Width of the Border (Meters)	Area of Houses and Parks (Square Meters)

2. Enter the data in the table above in a spreadsheet and plot the resulting graph.
3. Answer the following:
   (a) For what widths of the border is the required area for the houses and parks available?
   (b) Write a description or an equation showing what you did to determine the areas in the table for your guesses for the width of the border.
   (c) Generalize the description or equation in (b) to accommodate your teacher's guess of g meters for the width of the border. Write the resulting description, equation, or expression for the area in terms of the guess.
4. Compare the equations and descriptions other students found in 3(c), and determine if they are equivalent by doing the following:
   (a) Graph the different equations to see if the graphs are the same.
   (b) Use a computer algebra system to see if the equations or expressions simplify to the same values.
   (c) Use a computer algebra system to subtract the expressions found to see if the difference is 0.
5. Solve the problem.

Discussion and Extension

1. Find a problem in your algebra book, and use the techniques suggested in this activity to solve it.
2. What is the easiest method of solution for the chosen problem?

Solutions to "What's the Area?"

1. Values for a few different sample widths are shown.

Width of the Border (Meters)	Area of Houses and Parks (Square Meters)
25	1,377 500
26	1,372 704
27	1,367 916
28	1,363 136
29	1,358 364
30	1,353 600
31	1,348 844
32	1,344 096
33	1,339 356
34	1,334 624
35	1,329 900
36	1,325 184
37	1,320 476
38	1,315 776
39	1,311 084
40	1,306 400

2. The graph for the values in the sample table is shown.

3. (a) For certain widths between 30 and 31 meters
   (b) With the spreadsheet, the area is found using
   =(1500–2*A2)*(1000–2*A2),
   where A2 denotes the cell in which the width of the border is entered. (A2 as a variable in the spreadsheet fill-down mode changes.)
   (c) Area = $(1500 - 2g)(1000 - 2g)$.

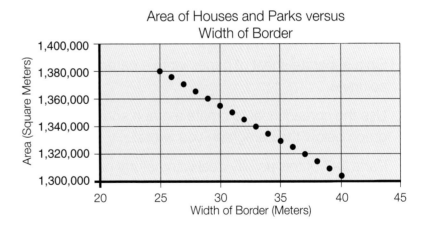

4. They are equivalent.

   (a) The graphs are the same.
   (b) The equations are the same.
   (c) The difference is 0.

5. The border is approximately 30.7 meters wide.

Discussion and Extension

1–2. The answer depends on the chosen problem.

When a Variable Is Not Single

Name _____

1. Use number line *a* below to mark the first ten natural numbers. Use number line *b* to mark the sequence of numbers whose values are three times the values of the numbers in the sequence on line *a*. Add any needed points on line *b* to indicate missing numbers. Draw arrows to complete the mapping diagram showing which number on line *a* is paired with which number on line *b*.

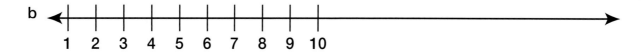

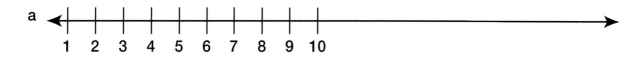

2. Describe with a recursive formula the pattern of the sequence that you drew on line *a*.
   _____

3. Describe with a recursive formula the pattern of the sequence you drew on line *b*.
   _____

4. Write an explicit formula of the form $f(x) =$ _____ that describes the pairing from line *a* to line *b*.
   _____

5. Create a spreadsheet that shows the functions mentioned in the steps above in this activity.

6. On a separate sheet of paper, write a paragraph describing how you constructed the spreadsheet columns.

7. Use the spreadsheet to create a graph showing the pairing from column A to column B. Sketch the graph below the paragraph you wrote.

8. Parts 1–7 of this activity deal with natural numbers. If you consider a domain of all the real numbers and the subset of real numbers greater than or equal to 1 and less than or equal to 50 as well as the graph that you created in part 7, what will be the set of images (real numbers on the vertical axis) of this subset of real numbers?

   _____
   _____
   _____

9. How would you depict the set of images on your graph? _____
   _____

When a Variable Is Not Single (continued)

Name _____

10. If you consider the set of real numbers that are greater than or equal to 100 and less than or equal to 1000 on the vertical axis of your graph, what set of preimages (real numbers on the horizontal axis) is paired with this set of numbers? _____

11. Write an inequality to represent the set of all numbers y such that y is greater than or equal to 100 and less than or equal to 1000. _____

12. Use the inequality and any properties of real numbers to find the set of preimages of the set in part 11.

13. What is the image of the set of numbers $237 < x \leq 450$, where x is a real number? _____

14. What is the preimage of the set of numbers $237 < x \leq 450$, where x is a real number? _____

Discussion and Extension

For the relationship (multiplying by 3) described in the activity, answer the following:

1. What is the image of the set of all even real numbers? _____
2. Describe the set of images of the multiples of 4. _____
3. Describe, in general, the set of images of the set of multiples of n. _____
4. What is the set of preimages that produced the set of images that are all the even real numbers?
5. What is the preimage of the number $0.\overline{3}$? _____
6. What is the image of the interval of real numbers $(0.5, \pi)$? _____

Navigating through Algebra in Grades 9–12

Solutions to "When a Variable Is Not Single"

1. The diagram follows:

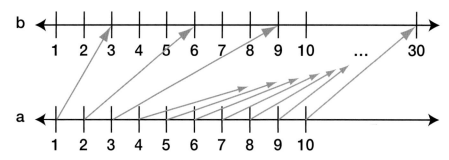

2. One recursive formula follows:

$$x_1 = 1$$
$$x_n = x_{n-1} + 1,$$

   where n is a natural number

3. One recursive formula follows:

$$y_1 = 3$$
$$y_n = y_{n-1} + 3,$$

   where n is a natural number

4. One explicit formula is $f(x) = 3x$, where x is a natural number less than or equal to 10.

5. One possible spreadsheet follows:

	A	B
1	1	3
2	2	6
3	3	9
4	4	12
5	5	15
6	6	18
7	7	21
8	8	24
9	9	27
10	10	30

6. Answers may vary. For example, column A was completed by entering 1 in cell A1, entering the formula =A1+1 in cell A2, and then using the fill-down command to complete the ten rows. For column B, 3 was entered in cell B1; the formula =B1+3 was entered in cell B2, and then the fill-down command was used.

7. The graph is shown in figure A1.

8. If the domain is the set of all real numbers, then the set of images of the set [1, 50], where the brackets are used to indicate that the set includes its endpoints, is the set [3, 150]. It is the set of real numbers y such that $3 \leq y \leq 150$.

Solutions to "When a Variable Is Not Single" (continued)

Fig. A1.

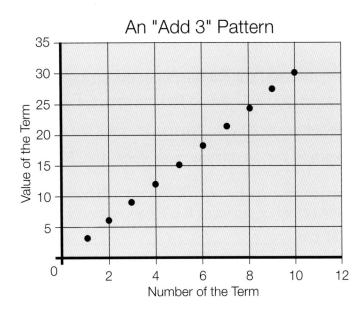

9. The set of images would be a solid segment connecting the dots of the graph in number 7.

10. The set of preimages is the set [100/3, 1000/3].

11. $100 \leq y \leq 1000$.

12. The solution might appear as follows:

$$100 \leq y \leq 1000$$
$$y = f(x) = 3x$$
$$100 \leq 3x \leq 1000$$
$$\frac{100}{3} \leq x \leq \frac{1000}{3}$$

13. The image is the set of numbers $711 < f(x) \leq 1350$.

14. The preimage is the set of numbers $79 < x \leq 150$.

Discussion and Extension

1. The image is the set of all multiples of 6.
2. The image is the set of all multiples of 12.
3. The image is the set of all real numbers $3n$, where n is a natural number.
4. The set of preimages is the set of all even multiples of 1/3.
5. The preimage is $0.\overline{1}$.
6. The image is the set $(1.5, 3\pi)$.

Navigating through Algebra in Grades 9–12

Tolerating a Bit

Name _____

As the quality control manager for a compact-disc manufacturer, you know that the desired circumference of a disc is 37 cm.

1. (a) To model the manufacturing process, cut out five paper circles (circular discs) with the desired circumference of 37 cm. To do so, you will have to find the radius, r.

   (b) By inspection, determine whether or not your five model discs are all alike.

   (c) Use a piece of string to measure the circumference of each disc. Record your measurements. _____

2. (a) Determine the smallest interval that contains all five measurements from part 1 as well as the desired circumference of 37 cm. _____

   (b) On the basis of your measurements, use number line C to represent the circumferences of the discs and to illustrate the *accuracy interval*, an interval on the number line showing the greatest and smallest circumference generated by the production process.

   (c) Express the accuracy interval as an inequality and by using set notation. _____

3. (a) Use the radius r, which corresponds to the desired circumference of a compact disc, to determine a *tolerance interval*, an interval of numbers about the radius indicating the greatest and smallest radii that will allow the manufacturer to have a circumference within the desired accuracy level, which corresponds to the accuracy interval you established in part 2b.

   (b) The second number line, below, labeled r representing radius, should be used to illustrate the tolerance interval.

   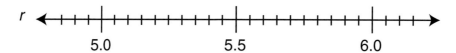

   (c) Express the tolerance interval using set notation. _____

4. As shown in figure A2, draw an arrow from several values for the radius of a compact disc on line r to the corresponding values for the circumference on line C. This type of model is a *mapping diagram* showing a function from r to C.

Adapted from Montana Council of Teachers of Mathematics, *Integrated Mathematics: A Modeling Approach with Technology*, Level 4 (Needham Heights, Mass.: Simon and Schuster Custom Publishing, 1997, pp. 134–36.

Tolerating a Bit (continued)

Name _____

Fig. A2.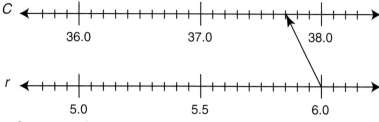

5. Compile the class data from part 1.

6. (a) Determine an accuracy interval for circumferences and a tolerance interval for radii that include all the class data. Express these intervals as inequalities. _____

   (b) Represent these intervals graphically using the number lines below.

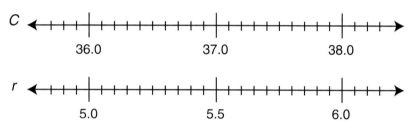

Discussion and Extension

1. Were all the model discs exactly the same size? _____ If not, what factors might explain the differences in size? _____

2. What function relates the radius of a compact disc to its circumference? _____

3. Describe the accuracy interval and the tolerance interval created for the class data. _____

4. How do your intervals from 2(a) and 3(a) in the exploration compare with the class's values? _____

5. In part 3 of the exploration, you determined the tolerance interval for the radius of your five compact discs using the accuracy interval for the resulting circumference. Describe this relationship using an if-then sentence. _____

6. Given any accuracy interval for a desired circumference C, how would you determine the corresponding tolerance interval for the radius r? _____

7. Describe how to represent an interval, unrelated to this problem, indicating each of the following sets of numbers:

   (a) All nonnegative real numbers _____ (b) All negative real numbers _____

Navigating through Algebra in Grades 9–12

Solutions to "Tolerating a Bit"

1. (a) Students create five paper circles that represent compact discs with a circumference of 37 cm.

   (b) Under normal circumstances, the discs will vary. Even in manufacturing, discs are not all exactly alike.

   (c) The appropriate number of significant digits should be determined by the calibration of the measuring instrument. The following sample data are reported to the nearest 0.1 cm: 36.8; 37.4; 37.1; 37.7; 37.4.

2. (a) The interval will vary with the students' measurements. With the sample data, the bounds of the accuracy interval are 36.8 and 37.7.

   (b) A graph of the accuracy interval for the sample data is shown.

   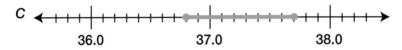

   (c) $36.8 \leq x \leq 37.7$ and x is an element of the set $\{x \mid 36.8 \leq x \leq 37.7$, where x is a real number$\}$.

3. (a) The tolerance interval for the sample data has bounds 5.9 and 6.0.

   (b) A graph showing the tolerance interval follows:

   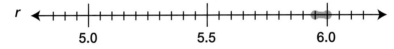

   (c) The radius r is an element of the set $\{r \mid 5.9 \leq r \leq 6.0$ and r is a real number$\}$.

4–6. The remainder of the answers will vary depending on students' data.

Discussion and Extension

1. Normally the model discs will not all be the same size because of discrepancies in measurement and in the construction of the circles.

2. The function is $C = 2\pi r$, or $f(r) = 2\pi r$.

3. The answers depend on the class data.

4. The answer depends on the class's data.

5. A sentence such as the following is appropriate: If the radius of the disc is in the tolerance interval [insert class interval], then the circumference of the disc is in the accuracy interval [insert class interval].

6. Use inequalities to find r. The determination might proceed as follows: Suppose C is in the interval $(C - a, C + a)$, then $C - a < 2\pi r < C + a$. Solve for r for the given value of a and C.

7. Students may come up with various schemes, but the following are customary:

   (a) $[0, +\infty)$

   (b) $(-\infty, 0)$

Tolerance and Accuracy

Name _____

1. The figure of an arbitrary linear function shows line l, the graph of a linear function $f(x)$. Using a geometry utility, create a drawing similar to the one shown. Your construction should meet the following conditions:

    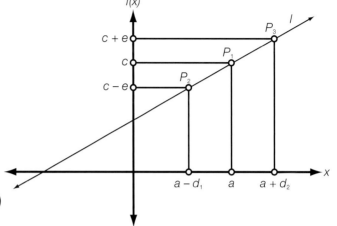

    (a) Line l is oblique to the $f(x)$-axis and represents a relationship between a desired measure c and its corresponding x-value, a.

    (b) The point with coordinates $(0, c - e)$ on the $f(x)$-axis is a movable point. The interval $(c - e, c + e)$ represents an accuracy interval for c.

    (c) The segment from point P_1 to the point with coordinates $(0, c)$ is perpendicular to the $f(x)$-axis. The segment from point P_1 to the point with coordinates $(a, 0)$ is perpendicular to the x-axis.

    (d) The point with coordinates $(0, c + e)$ is the reflection of the point with coordinates $(0, c - e)$ in the segment from point P_1 to the point with coordinates $(0, c)$.

    (e) The segment from the point with coordinates $(0, c - e)$ perpendicular to the $f(x)$-axis intersects l at P_2. The segment from point P_2 to the point with coordinates $(a - d_1, 0)$ is perpendicular to the x-axis.

    (f) The segment from the point with coordinates $(0, c + e)$ is perpendicular to the $f(x)$-axis and intersects l at P_3. The segment from point P_3 to the point with coordinates $(a + d_2, 0)$ is perpendicular to the x-axis. The interval $(a - d_1, a + d_2)$ represents the tolerance interval.

2. Measure the appropriate segments in order to determine the following distances:

    (a) Length of e _____

    (b) Length of d_1 _____

    (c) Length of d_2 _____

3. While moving the point with coordinates $(0, c - e)$ on the $f(x)$-axis, observe the values of the three distances listed in part 2, and record the values for several points.

    _____

4. (a) Use an inequality to describe the accuracy interval in terms of $f(x)$. _____

    (b) Use an inequality to describe the tolerance interval in terms of x. _____

Adapted from Montana Council of Teachers of Mathematics, *Integrated Mathematics: A Modeling Approach with Technology*, Level 4, vol. 2 (Needham Heights, Mass.: Simon and Schuster Custom Publishing, 1997, pp. 139–42

Tolerance and Accuracy (continued)

Name _____

Discussion and Extension

1. How does a change in e affect the values of d_1 and d_2? _____

2. Describe the effect that the size of the accuracy interval has on the size of the tolerance interval.

3. If point $P(x_1, 0)$ is on the line segment with endpoints $(a - d_1, 0)$ and $(a + d_2, 0)$, where would you expect the point with coordinates $(0, f(x_1))$ to be located? _____

4. In this case, once an accuracy interval is selected, how would you determine the corresponding tolerance interval? _____

5. Since d_1 and d_2 are equal, what mathematical concept may be used to describe the interval $(a - d_1, a + d_2)$? _____

6. What inequality can be used to express the interval $(a - d_1, a + d_1)$?

Solutions to "Tolerance and Accuracy"

1–2. Students' pictures should be similar to the following. (Numbers have been added as an example.)

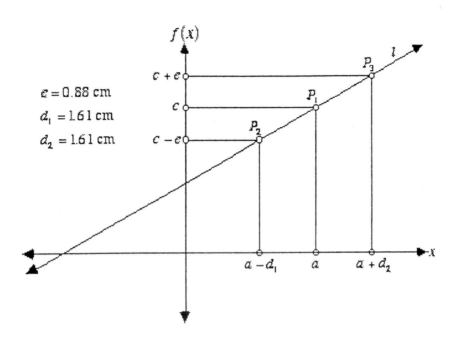

$e = 0.88$ cm
$d_1 = 1.61$ cm
$d_2 = 1.61$ cm

3. The following are sample data:

e	d_1	d_2
0.494	0.904	0.904
0.600	1.098	1.098
0.776	1.421	1.421
0.882	1.614	1.614
0.988	1.808	1.808
1.164	2.131	2.131
1.270	2.325	2.325
1.446	2.647	2.647
1.552	2.841	2.841
1.799	3.293	3.293

4. (a) The interval is $c - e < f(x) < c + e$.
   (b) The interval is $a - d_1 < x < a + d_2$.

Solutions to "Tolerance and Accuracy" (continued)

Discussion and Extension

1. As e increases, both d_1 and d_2 increase. As e decreases, both d_1 and d_2 decrease.

2. The larger the accuracy interval, the larger the corresponding tolerance interval. The smaller the accuracy interval, the smaller the corresponding tolerance interval.

3. Since the tolerance interval $(a - d_1, a + d_2)$ corresponds with the accuracy interval $(c - e, c + e)$, it follows that if $P(x_1, 0)$ is a point in the interval $(a - d_1, a + d_2)$, then $f(x_1)$ must be in the interval $(c - e, c + e)$.

4. In this case, because l is a strictly increasing line, one can determine the x-values that correspond to the $f(x)$-values that are the endpoints of the accuracy interval. These resulting x-values are the endpoints of the tolerance interval. *Note:* Although this process works on oblique lines, you may wish to caution students that it does not work in general.

5. An absolute value can be used.

6. $|x - a| < d_1$

Would You Work for Me?

Name _____

Would you work for me or for my sister with the following salary schemes? We will pay you $1 for your first day's work and $.50 for the second day's work. Each day after the second, your salary will be computed as follows:

$$\text{Tomorrow's salary} = \left(2\tfrac{1}{2}\right)(\text{Today's salary}) - \text{Yesterday's salary}$$

Furthermore, we don't like to bother with pennies but disagree on how they should be dealt with in computing your salary. As a the-glass-is-half-full optimist, I always round up to the nearest dime. My sister, however, as a the-glass-is-half-empty person, always rounds down to the nearest dime. Would you work for either of us? Why or why not?

1. After reading the statement of salary proposals, decide if you would work for either me or my sister. Write down your answer. _____
   _____

2. Complete the following chart:

Number of Day	Salary to Work for Me	Salary to Work for My Sister
1	$1.00	$1.00
2	$0.50	$0.50
3		
4		
5		
6		
7		
8		
9		

3. On the basis of the table you completed, would you stand by your decision? _____
   Explain why or why not. _____

4. From the table of daily salaries that you earn from me, what does the sequence become after six days? _____ Do you recognize it? _____

5. From the table of daily salaries that you earn from my sister, what does the sequence become after seven days? _____ Do you recognize it? _____

6. In my sister's salary scheme, what is the meaning of the salary on day 4? _____
   On day 5? _____

7. Is either salary scheme really realistic? _____

Navigating through Algebra in Grades 9–12

Would You Work for Me? (continued)

Name _____

Discussion and Extension

1. If you graphed the data for either salary scheme, what type of graph would you expect to obtain? _____ Why? _____
   _____

2. Describe some real-life situation in which one might encounter such schemes as those in the problem. _____

Solutions to "Would You Work for Me?"

1. Students will probably be divided over whether or not they will work under either salary scheme.
2. Some entries for the table are given.

Number of Day	Salary to Work for Me	Salary to Work for My Sister
1	$1.00	$1.00
2	$0.50	$0.50
3	$0.30	$0.20
4	$0.30	0
5	$0.50	–$0.20
6	$1.00	–$0.50
7	$2.00	–$1.10
8	$4.00	–$2.30
9	$8.00	–$4.70

3. The answer to standing by the decision will depend on what the student chose.
4. From day 6 on, it is apparent that the sequence is a geometric sequence of powers of 2.
5. From day 7 on, it is apparent that the sequence is a negative geometric sequence in which the next term, $n + 1$, is equal to $-(2n + \$0.10)$.
6. The salary for day 4 is 0, indicating that there is no salary. The salary for day 5 is negative, which could be interpreted to mean that you pay my sister to work.
7. Neither salary scheme is realistic.

Discussion and Extension

1. After completing the table, students would probably expect to get exponential graphs. Each graph is discrete.
2. Students' answers may vary, but chain letters or pyramid schemes for making money may be in the discussion.

The Devil and Daniel Webster

Name _____

The devil made a proposition to Daniel Webster. The devil proposed paying Daniel for services in the following way:

On the first day, I will pay you $1000 early in the morning. At the end of the day, you must pay me a commission of $100. Then we will determine your next day's salary and my commission according to the following rule: I will double what you have at the end of the day, but you must double the amount that you pay me. Will you work for me for a month?

1. After reading the salary proposal, decide if you would work if you were Daniel. Write down your answer. _____

2. Complete the following chart:

Number of Day	Salary for Daniel Webster	Commission
1	$1000	$100
2		
3		
4		
5		
6		
7		
8		
9		
10		
11		
⋮		
30		

3. On the basis of the table you completed, would you stand by your decision? _____
   Explain why or why not. _____

4. If you would not work for a month, for how many days would you work? _____

5. From reading the problem, what type of curve would you expect your salary data to generate?
   _____

6. From reading the problem, what type of curve would you expect the commission data to generate?
   _____

7. Is the salary scheme realistic? _____

The Devil and Daniel Webster (continued)

Name _____

Discussion and Extension

1. If you graphed your salary data, what type of graph would you expect to obtain? _____
   Why? _____
   _____

2. In your own words, compare the graphs of the salary data and the commission data. _____
   _____
   _____

Solutions to "The Devil and Daniel Webster"

1. The answers will depend on the student. The outcome is not apparent in the beginning.
2. Part of the table is shown below.

Number of Day	Salary for Daniel Webster	Commission
1	$1,000	$100
2	$1,800	$200
3	$3,200	$400
4	$5,600	$800
5	$9,600	$1,600
6	$16,000	$3,200
7	$25,600	$6,400
8	$38,400	$12,800
9	$51,200	$25,600
10	$51,200	$51,200
11	0	$102,400
⋮	⋮	⋮
30	-1.02×10^{12}	5.3687×10^{10}

3. The answer depends on what the student decided early on. The student definitely will not want to work long with this commission.
4. Working nine days is reasonable with the given scheme.
5. Most will expect an exponential curve.
6. Most students will expect an exponential curve.
7. The salary scheme is not realistic.

Discussion and Extension

1. Most will expect the graph to be an exponential curve. It is not at all clear that they will expect the curve that they get. (See the graph at the right.)
2. The graphs are both of the exponential type. The commission curve grows "faster" than the salary curve.

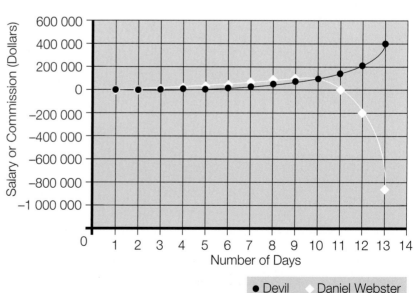

Iterating to Find the Square Root of 2

1. Pick some number that you know is greater than $\sqrt{2}$ as the first guess in determining this root. _____

2. Use the following formula for the "NEXT" guess:

$$\text{NEXT} = \frac{\text{CURRENT GUESS} + \dfrac{2}{\text{CURRENT GUESS}}}{2}$$

   Your guess times the quotient of 2 divided by your guess equals 2:

$$\text{GUESS} \times \frac{2}{\text{GUESS}} = 2$$

   If your guess is less than $\sqrt{2}$, then the quotient 2/GUESS will be greater than $\sqrt{2}$. The formula averages GUESS and 2/GUESS to yield NEXT, the next (better) guess.

   What is your second guess? _____

3. To use this algorithm on a calculator, do the following:

   (a) Store your guess in A: On a TI-92 enter your guess and press $\boxed{\text{STO A}}$.

   (b) On the command line of a TI-92, enter (A + (2/A))/2, and press $\boxed{\text{STO A}}$.

   (c) Press $\boxed{\text{ENTER}}$ several times. Write the sequence of values that you obtain.

   _____ _____ _____ _____
   _____ _____ _____ _____
   _____ _____ _____ _____

   (d) Find $\sqrt{2}$ with your calculator. _____

   (e) How accurate was the iterative algorithm you used? _____
   _____

Discussion and Extension

1. Is there a way to represent this iterative process graphically in order to see that it really does find $\sqrt{2}$? _____ Explain any ways you discovered. _____
_____

2. Is there an algebraic argument for convincing someone that the iterative process used really does keep getting closer to $\sqrt{2}$? _____ State your argument. _____
_____

Navigating through Algebra in Grades 9–12

Iterating to Find the Square Root of 2 (continued)

3. How many iterations would be required to get accuracy to one hundred decimal places in the estimate of $\sqrt{2}$?
   _____

4. If the process converges to $\sqrt{2}$, how "fast" does it converge? That is, how does the error change from iteration to iteration? _____
   _____

Solutions to "Iterating to Find the Square Root of 2"

1. Suppose the initial guess is 2. Then the following solutions apply.
2. 1.5
3. (c) 1.41666667, 1.41421569, 1.41421356, 1.41421356, 1.41421356, 1.41421356, 1.41421356, 1.41421356, 1.41421356, 1.41421356, 1.41421356, 1.41421356, 1.41421356, 1.41421356, 1.41421356, 1.41421356, 1.41421356, 1.41421356, 1.41421356, 1.41421356,
   (d) 1.41421356237
   (e) The algorithm is very accurate.

Discussion and Extension

1. Yes. There are several ways to do this. One is to use a web plot, as shown in chapter 3. Another is to plot the points as a sequence and compare the curve of the sequence to the line $y = \sqrt{2}$.

2. Yes. If the guess is correct, the solution to the equation is $\sqrt{2}$. If the guess is exactly equal to $\sqrt{2}$, then 2/GUESS = GUESS. But if the guess is less than $\sqrt{2}$, then 2/GUESS > $\sqrt{2}$, and vice versa. The average of GUESS and 2/GUESS lies between those values and hence is closer to $\sqrt{2}$. Each iteration gets progressively closer to $\sqrt{2}$.

3. With seven iterations, well over one hundred–decimal-place accuracy is achieved. (The exact number of places depends on the original guess.)

4. The process is extremely accurate. The error changes very little after the first five iterations.

Representing the Solution Process by Graphing

Name _____

1. Examine the solution to the following equation:

$$5(3 - x) + 4 = 2x - 9$$
$$15 - 5x + 4 = 2x - 9$$
$$19 - 5x = 2x - 9$$
$$19 + 9 = 2x + 5x$$
$$28 = 7x$$
$$4 = x$$

   Think of each side of each equation as a separate equation to be graphed (written here as a graphing calculator might show it). Make a list of all the separate functions in the table below.

Function	Expression
y_1	$5(3 - x) + 4$
y_2	$2x - 9$
y_3	$15 - 5x + 4$
y_4	
y_5	
y_6	
y_7	
y_8	
y_9	
y_{10}	
y_{11}	
y_{12}	

2. What do you think all the functions y_1 through y_{12}, considered in pairs, have in common that will be apparent when you draw the graphs? _____

3. Use your graphing utility to sketch the pairs of functions, y_1 and y_2 through y_{11} and y_{12}. What do all the pairs of lines have in common? _____

_____

Navigating through Algebra in Grades 9–12

Representing the Solution Process by Graphing (continued)

Name _____

4. Try the same exercise as above on an equation of your choice.

    (a) List the equation here. _____

    (b) In the space below, solve the equation.

    (c) Make a table like that in part 1 showing the functions.

    (d) Sketch the graphs of the pairs of functions on graph paper.

    (e) What do all the graphs have in common? _____

Discussion and Extension

1. Explain whether you think the commonality in the pairs of functions will always occur. _____

2. Each pair of functions used in the exercise is a system of linear equations. What do you think is necessary in order for two systems of equations to be equivalent? _____

Navigating through Algebra in Grades 9–12

Representing the Solution Process by Graphing (continued)

Name _____

3. Consider the original equation:

$$5(3 - x) + 4 = 2x - 9$$

Write a new equation from the one given with all the terms moved to the left side of the equation so that the right side is 0. If you graphed the functions y_1 and y_2, where would the lines obtained intersect? _____ What is the meaning of this point of intersection?

_____

_____

4. Consider the equation that follows:

$$\frac{x^2 - 9}{x - 3} = 0$$

(a) Use the methods suggested in this worksheet to find a solution.

(b) Is the solution found really a solution to the original equation? _____

Explain why or why not. _____

_____

_____

Solutions to "Representing the Solution Process by Graphing"

1.

Function	Expression
y_1	$5(3 - x) + 4$
y_2	$2x - 9$
y_3	$15 - 5x + 4$
y_4	$2x - 9$
y_5	$19 - 5x$
y_6	$2x - 9$
y_7	$19 + 9$
y_8	$2x + 5x$
y_9	28
y_{10}	$7x$
y_{11}	4
y_{12}	x

2. Students might conjecture that all the pairs of functions have the same x–coordinate of their points of intersection.

3. The graph is shown at the right. All the pairs of functions have the same x-coordinate of their points of intersection.

4. Answers will vary according to the equations chosen.

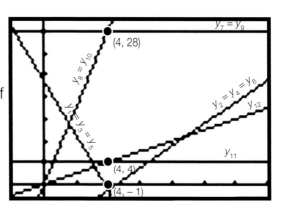

Discussion and Extension

1. The pairs of equations will always have a common point of intersection when the solutions involve linear expressions.

2. The system must have the same solution.

3. $28 - 7x = 0$. The lines intersect at the point with coordinates (4, 0). This point gives the solution to the original equation.

4. (a) The graphs appear to intersect at the point with coordinates (3, 6).

   (b) The solution is not really a solution because x cannot be equal to 3. If $x = 3$, then the left side of the equation contains a division by 0.

Navigating through Algebra in Grades 9–12

Families of Functions

Name _____

In this activity, you will work with a set of real data and first find an equation to model the data. After that, you will examine different aspects of families of functions. Consider the median salary in dollars for women and men in the table below, taken from the U.S. Census data (www.census.gov/income/p13.txt).

Median Salaries for Workers Twenty-five Years or Older according to Gender

Year	Median Salary: Women	Median Salary: Men
1991	$11,580	$23,686
1992	$11,922	$23,894
1993	$12,234	$24,605
1994	$12,766	$25,465
1995	$13,821	$26,346
1996	$14,682	$27,248
1997	$15,573	$28,919

1. Enter the data for the years and women's salaries in a spreadsheet or graphing calculator.

2. Make a graph depicting the data, using years on the horizontal axis.

3. Find the equation of a line that approximates the data. _____

4. How would the median salaries have changed if all them had been $500 more? _____

5. Add a new column to the spreadsheet or new data into the graphing calculator that shows salaries with the increment suggested in part 4. How do you expect the scatterplot of women's salaries versus time to change?

6. Draw the scatterplot, and determine an equation for the new data. _____

   How does this equation compare with the first one? _____

7. Look at the salary data in the first column and in the second column. Look at the equation that you obtained in part 3. How can you approximate the data in the men's column from the data in the women's column? _____

8. Sketch a mapping diagram showing the time period and the first set of data.

Families of Functions (continued)

Name _____

9. Change the mapping diagram to include a third horizontal line showing how to obtain the second set of data from the first.

10. Describe mathematically the process that is being used both in the mapping diagram and on the spreadsheet. _____
_____

Discussion and Extension

1. Composition of functions is found in all areas of daily life. Describe how putting money in a pop machine and then receiving change and a soda could be a composition of functions. What are the domains and ranges of the functions described? _____
_____
_____
_____

2. Describe the advantages and disadvantages of using a mapping diagram versus using a graph.
_____
_____
_____

Navigating through Algebra in Grades 9–12

Solutions to "Families of Functions"

Name _____

1–2. A graph depicting the data is given below.

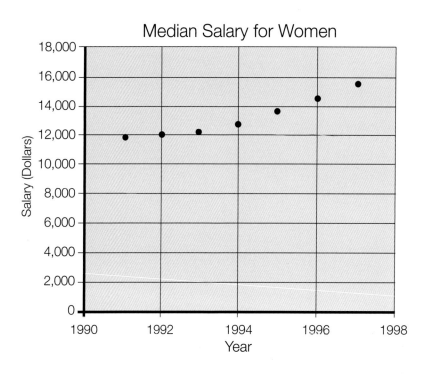

3. One possible equation is $y = 681x - 1{,}344{,}500$. If the years are labeled as 1, 2, 3, ..., 7, then an equation might be $y = 681x + 10\,498$. The latter might be the preferred equation, but to use it, students will have to understand that 1990 is subtracted from each year given to make the graph of the data easier to handle. In terms of functions, the first equation is being transformed by a translation to make the first data point closer to the origin of the graph.

4. If the salaries had been $500 more in each period, then the median would have been $500 greater.

5. Each dot in the graph should be 500 units higher.

6. If the latter equation is used, the new equation should be approximately $y = 681x + 10{,}998$. The equation should differ by $500.

7. Because each value in the men's salary column is approximately $12,512 greater than the corresponding value in the women's salary column, that amount could be added to the corresponding values in the women's column.

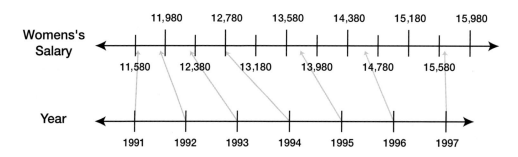

78 Navigating through Algebra in Grades 9–12

Solutions to "Families of Functions" (continued)

8. The mapping diagram follows:

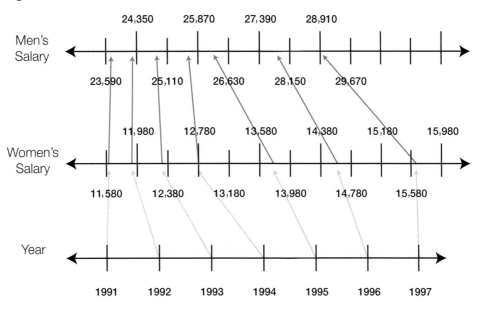

9. The mapping diagram follows:

10. In each case, there is a composition of functions. To write one function that would accomplish the same thing, one could use part 9 and draw an arrow from the lower line to the upper line to show the new function.

Discussion and Extension

1. Answers will vary. The teacher should listen to the students to see what their functions are.

2. A mapping diagram is cumbersome to construct but shows the relation between the domain and the range of a function very well. A graph shows a more "dynamic relation" between the domain and the range and may allow predictions about data to be made more easily.

Operating on Functions

Name _____

1. Consider the functions $f(x) = x^2$ and $g(x) = x - 1$.

2. For each function, create a spreadsheet with stepped values of x and the corresponding values of the function.

3. Using a graphing calculator or a graphing utility on a computer, graph $f(x)/g(x)$ and $g(x)/f(x)$ on separate graphs using connected scatterplots.

4. Where are the functions undefined? _____

   Why? _____

5. Would a finer grid of stepped values of x have made the graphs be defined? _____

   Why or why not? _____

6. What are the domains of the two functions $f(x)$ and $g(x)$? _____
   _____

7. What are the domains of the two functions $f(x)/g(x)$ and $g(x)/f(x)$?

Discussion and Extension

1. A rancher would like to create a rectangular corral with an area of 150 m², using the least amount of fencing material.

   (a) Write a function that describes the perimeter of the rectangle in terms of the width of the rectangle.
   _____

   (b) Describe the domain. _____

   (c) What does the domain mean in the given context? _____

   (d) Graph the function.

2. (a) Graph the following using a graphing calculator:

   $$f(x) = \frac{x^2 - 5}{x}$$

   (b) What is the domain of $f(x)$? _____

   (c) For the graph on the graphing calculator, press ZOOMOUT several times with the same center. What happens to the graph? _____

   (d) What is the equation of the line that you see after zooming out several times? _____

   (e) Use a computer algebra system to divide $x^2 - 5$ by x. What is the quotient? _____

   What is the remainder? _____

Operating on Functions (continued)

Name _____

(f) How do the quotient in part 2(e) and the equation in part 2(d) compare? _____

(g) Consider the remainder found in part 2(e) in fractional form. What happens to the value of the fraction as x becomes extremely large? _____

(h) What is the mathematical relation of the line found in part 2(d) to the original function? Write an explanation of this relation. _____

_____

Solutions to "Operating on Functions"

1–2. A spreadsheet for both functions is shown at the right.
3. The scatterplots follow.

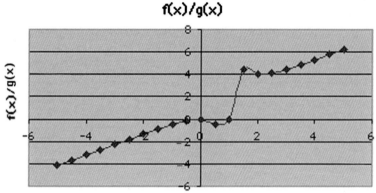

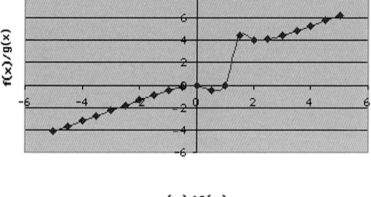

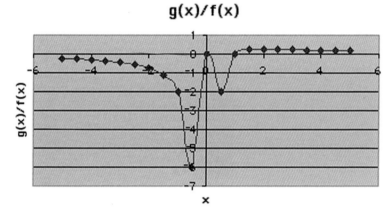

	A	B	C
1	x	f(x)=x^2	g(x)=x−1
2	−5	25	−6
3	−4.5	20.25	−5.5
4	−4	16	−5
5	−3.5	12.25	−4.5
6	−3	9	−4
7	−2.5	6.25	−3.5
8	−2	4	−3
9	−1.5	2.25	−2.5
10	−1	1	−2
11	−0.5	0.25	−1.5
12	0	0	−1
13	0.5	0.25	−0.5
14	1	1	0
15	1.5	2.25	0.5
16	2	4	1
17	2.5	6.25	1.5
18	3	9	2
19	3.5	12.25	2.5
20	4	16	3
21	4.5	20.25	3.5
22	5	25	4

4. From the graphs as shown, it is not clear where the functions are undefined. However, the spreadsheet shows that $f(x)/g(x)$ is undefined when $x = 1$. The second function, $g(x)/f(x)$, is undefined when $x = 0$.

5. It depends on the grid, but a finer grid would not have made the graphs defined.

6. The domain of $f(x)$ is the set of all real numbers, as is the domain of $g(x)$.

7. The domain of $f(x)/g(x)$ is the set of all real numbers except 1. The domain of $g(x)/f(x)$ is the set of all real numbers except 0.

Discussion and Extension

1. (a) If the width is w, then the perimeter P can be written as $P(w) = 2w + 150/w$.

   (b) The domain of $P(w)$ is the set of all positive real numbers.

   (c) The width of the corral must be a positive number. In a real situation, there may be other restrictions.

Solutions to "Operating on Functions" (continued)

(d) The graph follows:

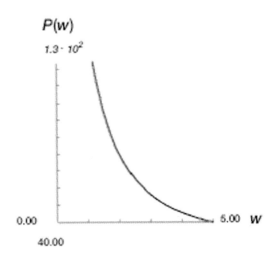

2. (a) The graph follows:

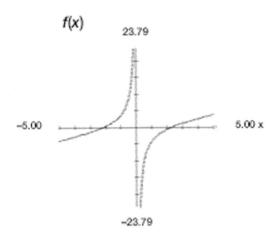

(b) The domain is the set of all real numbers but 0.

(c) If one continues to zoom out, the graph becomes a straight line.

(d) The equation is $y = x$.

(e) The quotient is x. The remainder is -5.

(f) The quotient is the right side of the equation of the line.

(g) The remainder in fractional form is $5/x$. As x gets extremely large, the value of the fraction goes to 0.

(h) The line is the limit of the function as x gets extremely large or small.

References

Brieske, Thomas J. "Mapping Diagrams and the Graph of $y = \sin 1/x$." *Mathematics Teacher* 73 (April 1980): 275–78.

Coxford, Arthur F., James T. Fey, Christian R. Hirsch, Harold L. Schoen, Gail Burrill, Eric W. Hart, Ann E. Watkins, Mary Jo Messenger, and Beth Ritsema. *Course 1, Part A, Contemporary Mathematics in Context.* Core-Plus Mathematics Project. Chicago: Everyday Learning, 1997.

Demana, Franklin, and Joan Leitzel. "Establishing Fundamental Concepts through Numerical Problem Solving." In *The Ideas of Algebra, K–12*, 1988 Yearbook of the National Council of Teachers of Mathematics (NCTM), edited by Arthur F. Coxford. Reston, Va: NCTM, 1988.

Devaney, Robert L. "Putting Chaos into the Classroom." In *Discrete Mathematics across the Curriculum, K–12*, 1991 Yearbook of the National Council of Teachers of Mathematics (NCTM), edited by Margaret J. Kenney, pp. 184–94. Reston, Va.: NCTM, 1991.

Hahn, Alexander J. *Basic Calculus from Archimedes to Newton to Its Role in Science.* New York: Springer-Verlag, 1998.

House, Peggy. *Mission Mathematics, Grades 9–12*. Reston, Va.: National Council of Teachers of Mathematics, 1997.

Kowalczyk, Robert E., and Adam O. Hausknecht. TeMATH, Tools for Exploring Mathematics. Version 1.5. Pacific Grove, Calif.: Brooks/Cole Publishing Co., 1993.

LaTorre, Donald R., John Kenelly, Iris Fetta, Cynthia Harris, and Laurel Carpenter. *Calculus Concepts: An Informal Approach to the Mathematics of Change.* Lexington, Mass.: D. C. Heath & Co., 1995.

Montana Council of Teachers of Mathematics. *Integrated Mathematics: A Modeling Approach Using Technology, Level 4*. Needham Heights, Mass.: Simon & Schuster Custom Publishing Co., 1997.

———. *Integrated Mathematics: A Modeling Approach Using Technology, Level 6*. Needham Heights, Mass.: Simon & Schuster Custom Publishing Co., 1998.

National Council of Teachers of Mathematics (NCTM). *Historical Topics for the Mathematics Classroom.* Thirty-first Yearbook of the National Council of Teachers of Mathematics. Washington, D.C.: NCTM, 1969.

———. *Principles and Standards for School Mathematics.* Reston, Va.: NCTM, 2000.

Rosnick, Peter. "Some Misconceptions concerning the Concept of Variable." *Mathematics Teacher* 74 (September 1981): 418–20. Reprint in *Algebraic Thinking, Grades K–12: Readings from NCTM's School-Based Journals and Other Publications*, edited by Barbara Moses, pp. 313–15. Reston, Va.: National Council of Teachers of Mathematics, 1999.

Sherin, Miriam G., Edith P. Mendez, and David A. Louis. "Talking about Math Talk." In *Learning Mathematics for a New Century*, 2000 Yearbook of the National Council of Teachers of Mathematics (NCTM), edited by Maurice J. Burke, pp. 188–96. Reston, Va.: NCTM, 2000.

Usiskin, Zalman. "Conceptions of School Algebra and Uses of Variables." In *The Ideas of Algebra, K–12*, 1988 Yearbook of the National Council of Teachers of Mathematics (NCTM), edited by Arthur F. Coxford, pp. 8–19. Reston, Va.: NCTM, 1988.

Wagner, Clifford. "Hammurabi's Calculator." In *The Teaching and Learning of Algorithms in School Mathematics*, 1998 Yearbook of the National Council of Teachers of Mathematics (NCTM), edited by Lorna J. Morrow, pp. 86–90. Reston, Va.: NCTM, 1998.

Wagner, Sigrid. "What Are These Things Called Variables?" *Mathematics Teacher* 76 (October 1983): pp. 474–79. Reprint in *Algebraic Thinking, Grades K–12: Readings from NCTM's School-Based Journals and Other Publications*, edited by Barbara Moses, pp. 316–20. Reston, Va.: National Council of Teachers of Mathematics, 1999.

Suggested Reading

Beaton, Albert E., Michael O. Martin, Ina V. S. Mullis, Eugenio J. Gonzalez, Teresa A. Smith, and Dana L. Kelly. *Mathematics Achievement in the Middle School Years: IEA's Third International Mathematics and Science Study (TIMSS)*. Chestnut Hill, Mass.: TIMSS International Study Center, Boston College, 1996.

Bouma, Carol A. "Design Your Own City: A Discrete Mathematics Project for High School Students." In *Discrete Mathematics across the Curriculum, K–12*, edited by Margaret J. Kenney, pp. 235–45. Reston, Va.: National Council of Teachers of Mathematics, 1991.

National Council of Teachers of Mathematics. Figure This! Available online at www.figurethis.org.

Philipp, Randolph A. "The Many Uses of Algebraic Variables." *Mathematics Teacher* 85 (October 1992): 557–61. Reprint in *Algebraic Thinking, Grades K–12: Readings from NCTM's School-Based Journals and Other Publications*, edited by Barbara Moses, pp. 157–62. Reston, Va.: National Council of Teachers of Mathematics, 1999.

Schoenfeld, Alan H., and Abraham Arcavi. "On the Meaning of Variable." *Mathematics Teacher* 81 (September 1988): 420–27. Reprint in *Algebraic Thinking, Grades K–12: Readings from NCTM's School-Based Journals and Other Publications*, edited by Barbara Moses, pp. 150–56. Reston, Va.: National Council of Teachers of Mathematics, 1999.

U.S. Department of Education. *Introduction to TIMSS: The Third International Mathematics and Science Study*. Washington, D.C.: U.S. Department of Education, September 1997.

Yerushalmy, Michal, and Shoshana Gilead. "Solving Equations in a Technological Environment." *Mathematics Teacher* 90 (February 1997): 156–62. Reprint in *Algebraic Thinking, Grades K–12: Readings from NCTM's School-Based Journals and Other Publications*, edited by Barbara Moses, pp. 268–74. Reston, Va.: National Council of Teachers of Mathematics, 1999.